Anticipation Science

Volume 2

Anticipation Science encompasses natural, formal, and social systems that intentionally or unintentionally use ideas of a future to act in the present, with a broad focus on humans, institutions, and human-designed systems. Our aim is to enhance the repertoire of resources for developing ideas of the future, and for expanding and deepening the ability to use the future. Some questions that the Series intends to address are the following: When does anticipation occur in behavior and life? Which types of anticipation can be distinguished? Which properties of our environment change the pertinence of different types of anticipation? Which structures and processes are necessary for anticipatory action? Which is the behavioral impact of anticipation? How can anticipation be modeled?

The series is interested in receiving book proposals that:

- are aimed at an academic audience of graduate level and up
- combine applied and/or theoretical and/or philosophical studies with work especially from disciplines within the human and social sciences broadly conceived.

More information about this series at http://www.springer.com/series/15713

A. H. Louie

Intangible Life

Functorial Connections in Relational Biology

Springer

A. H. Louie
Ottawa, Ontario
Canada

ISSN 2522-039X ISSN 2522-0403 (electronic)
Anticipation Science
ISBN 978-3-319-65408-9 ISBN 978-3-319-65409-6 (eBook)
DOI 10.1007/978-3-319-65409-6

Library of Congress Control Number: 2017947857

Printed on acid-free paper

This Springer imprint is published by Springer Nature
The registered company is Springer International Publishing AG
The registered company address is: Gewerbestrasse 11, 6330 Cham, Switzerland

故有之以為利無之以為用

[Therefore profit comes from what is tangible,
Usefulness from what is intangible.]

—Lao Tse (6th century BC)
Tao Te Ching
Chapter 11

To my companions
in Purgatory:
Keep swinging.

Nota bene

Prerequisites

Welcome to the *tertia opus* of our exploratory journey in relational biology! My two previous books

More Than Life Itself:
A Synthetic Continuation in Relational Biology

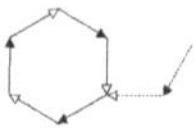

The Reflection of Life:
Functional Entailment and Imminence in Relational Biology

were published in 2009 and 2013, respectively. Since I shall be referring to these two books many times, henceforth the canonical symbols *ML* and *RL* will be used in their stead. In this present volume, I shall adopt the notation and terminology and draw upon results from *ML* and *RL*. When various topics are encountered, I shall when appropriate refer the reader to relevant passages in *ML* and *RL* for further exploration; the notation '*ML*: 5.13', for example, refers to Section 5.13 (in Chapter 5) of *More Than Life Itself.*

I assume the reader is already familiar with the premises of the Rashevsky–Rosen school of relational biology, as explicated in *ML*. I enlist all the suppositions made in *RL*: Nota bene (which recursively procures their predecessors in *ML*: Nota bene) and include them as prerequisites for continuing our journey in relational biology. In particular, the reader should know what (M,R)-networks and (M,R)-systems are, be able to distinguish between sequential and hierarchical compositions of mappings (*ML*: 5.13, 5.14) and to identify Aristotle's four causes as components of a mapping in its 'solid-headed' + 'hollow-headed' arrow notation (*ML*: 5.4–5.11), and have already understood the following statements.

Definition (*RL*: 6.10) The entailment of a material cause is called *material entailment.*

Definition (*ML*: 5.15; *RL*: 6.14) The entailment of an efficient cause is called *functional entailment.*

Definition (*ML*: 6.23; *RL*: 7.2) A natural system is *closed to efficient causation* if its every efficient cause is (functionally) entailed within the system.

Postulate of Life (*ML*: 11.28; *RL*: 8.30) A natural system is an *organism* if and only if it realizes an (M,R)-system.

Fundamental Theorem of Relational Biology (*ML*: 11.29; *RL*: 7.4) *A natural system is an organism if and only if it is closed to efficient causation.*

The Exordium of *RL* (pp. xvii–xxxii) is a concise review of relational biology.

As per the custom established in *ML* and *RL*, in this book I assume that the reader is versed in the basic facts of *naive set theory*, as presented, for example, in Halmos [1960]. Category theory is the metalanguage of relational biology, and I also suppose that the reader is conversant with this branch of abstract algebra. The definitive reference remains Mac Lane [1997]. A summary of those category-theoretic concepts that appear in my expositions may also be found in the Appendix of *ML* (pp. 329–372).

Throughout this book, I shall occasionally (re)define some mathematical concepts found in *ML* and *RL*. As an illustrative example, the Prolegomenon contains another take on an introduction to category theory. The repetition is often done for clarity of notations (especially for those non-standardized ones) and sometimes for alternative descriptions of concepts. It is also my mimicry of completeness (by placing definitive passages in the contextual flow of the narrative) and my attempt at consistence (by avoiding the appearance of a mathematical entity before its (re) definition).

Since I am assuming the reader is acquainted with relational biology and its mathematical metalanguage, I have relaxed the requirement of a prescribed sequential order in this monograph's presentation. I shall anticipate things and use terms that are yet undefined, without in situ complete discussions. The suspension would disappear when the preparation has sufficiently advanced, and thence resolution would follow. The inherent complexity of the subject dictates that at times it simply cannot be helped that some concepts depend on subjects that have come to pass as well as objects yet to come.

Praefatio
Nisi mittam manum meam in latus eius, non credam

> The *intangible* represents the real power of the universe. It is the seed of the tangible.
>
> —Bruce Lee (1940–1973)
> *Jeet Kune Do* (posthumously published in 1997)
> Part 6: Beyond System
> —The Ultimate Source of Jeet Kune Do

More Than Life Itself (*ML*, Louie 2009) explored the *epistemology* of life. *The Reflection of Life* (*RL*, Louie 2013) dealt with the *ontogeny* of life. This 'Opus III', the book that you are reading, emphasizes the *intangibility* of life.

The roots of the Medieval Latin word *intangibilis* are *in* 'without/not' + *tangere* 'touch', 'feel (in the sense of examine or search by touch)', 'handle'. So the original meaning of 'intangible' is 'unable to be touched', which neither comes with a judgment nor is burdened with negative implications. In our world overwhelmed by materialism (in all its senses), however, something not perceivable by touch, that has no material form or physical presence, is deemed undesirable. The pure meaning of 'intangible' has therefore been contaminated with sinister connotations of unintelligible, unable to be grasped mentally, having no value in itself, worthless, vague, indefinite, indefinable, obscure, unclear, elusive, etc. Fallen under the same spellbound bias are its similarly devolved synonyms: untouchable, impalpable, immaterial, insubstantial, etc.

It is therefore gratifyingly delightful that in 2003, the United Nations Educational, Scientific and Cultural Organization (UNESCO) adopted the *Convention for the Safeguarding of the Intangible Cultural Heritage* [UNESCO document 132540]. Thereby, in one broad stroke, UNESCO restored the rightful place that 'intangibles' ought to occupy in the world. The Convention's Article 1, paragraph 1, states:

> The "intangible cultural heritage" means the practices, representations, expressions, knowledge, skills — as well as the instruments, objects, artefacts, and cultural spaces associated therewith — that communities, groups and, in some cases, individuals recognize as part of their cultural heritage. This intangible cultural heritage, transmitted from generation to generation, is constantly recreated by communities and groups in response to their environment, their interaction with nature, and their history, and provides them with a sense of identity and continuity, thus promoting respect for cultural diversity and human creativity.

Since then, more than 300 elements have been inscribed onto the Representative List of the Intangible Cultural Heritage of Humanity, for example, Cantonese opera, falconry, tango, and Turkish coffee tradition. The intangibles of cultural heritage are the practices, representations, expressions, knowledge, skills; they dictate (and are manifested in) the associated tangible parts: the instruments, objects, artefacts, and cultural spaces.

The United Nations World Commission on Culture and Development had published in 1995 the report *Our Creative Diversity* [UNESCO document 101651, summary document 105586]. It had been a first step towards the ultimate adoption of the 2003 UNESCO Convention. Chapter 7 of the report, "Cultural Heritage for Development", contained this paragraph:

> **Broader visions are needed**
>
> Development presents new challenges for heritage conservation. Not only is there a huge gap between means and ends, but our definitions are still too narrow. They are biased towards the elite, the monumental, the literate, and the ceremonial. There is a need to reassess such conceptions as well as to develop better methods of identifying and interpreting our heritage. It is essential to understand the values and aspirations that drove its makers, without which an object is torn from its context and cannot be given its proper meaning. The tangible can only be interpreted through the intangible.

Amen. One sees the conceptual confluence with relational biology.

Function dictates structure. "The tangible can only be interpreted through the intangible". As we, practitioners of relational biology, have maintained, we are not denying that an underlying material basis is needed and from which *some* information on living systems may be derived. Cells and organisms *are*, after all, material systems, and molecules of the same chemical compound are identical to one another whether they originate in a living system or a nonliving system ('animal, vegetable, or mineral'), or, indeed, when artificially synthesized. The real *nature* of living systems, however, is conveyed not by their material basis but by their inherent processes. Physicochemical structures may imply functions; but for biological systems, the more important entailment is the converse: physicochemical structures are manifestations of functions.

Relational biology is *not* vitalism in disguise. *Vitalism* is, indeed, more an *abdication* of theory than a theory, its *modus operandi* being an assertion that the origin and phenomena of life are dependent on forces distinct from purely chemical or physical forces. This assertion is nothing but a euphemism for the resignation that the basic forces of biology, as yet irreducible to chemistry and physics, are thus known unknowns. *Reductionism*, on the flip side, is the other polar hyperbole, referring all biological properties ultimately back to the constituent molecules; this irrevocably welds biology to chemistry and to physics. Physics is always claimed to be the *universal* science, its subject matter comprising the study of *all* matter and its interactions. The antagonists, vitalists, and reductionists all, arguably, do agree in principle that scientific operating principles have to be tangible to be real. There is also concurrence (if only grudgingly admitted by at least the more honest reductionists) that our current understanding of physical and chemical is not sufficient to explain life. The difference lies in the two opposing attempts at resolution. Vitalists invoke *élan vital* (or one of its many guises, as "vital impetus", "life force", and so forth) as an additional axiom. A more enlightened reductionist such as Schrödinger may call for "new laws to be expected in an organism", "a new type of physical law", but remaining "new principle not alien to physics". A die-hard reductionist would presumptuously declare that the toolkit of *contemporary* physics suffices, and any perceived inability in complete explication of life is a temporary setback that they shall overcome. The irony is that even in classical physics, fields (e.g. electromagnetic fields, gravitational fields)—intangible and insubstantial—are well-accepted as part of the fabric of the universe. Indeed, it may even be argued that interacting fields are the only 'real' things in the universe, and matter is simply the momentary manifestations of fields as particles. Only the reductionist–biologist remains adamant in the doctrine of materialism (dialectal or otherwise) that "Seeing's believing, but feeling's the truth". To wit, they may reluctantly allow, say, an organism to react *as a physicochemical system* in an electromagnetic field, but any kind of mention of uniquely biologically generated fields is anathema to them.

To be sure, the reductionistic approach seems inevitable. This is because biology, ever since antiquity when there was not yet a subject called 'biology', has traditionally been a descriptive science, based on physical observations and manipulations. Modern biology evolved to exploit tools developed in chemistry, physics, technology, and engineering. The cell theory and the germ theory, for example, could never have been formulated without the availability of microscopes, instruments predicated on advancements in optics. Likewise, cell biology, biochemistry, and molecular biology presuppose advancements in atomic science and computer technology, spawning instruments such as electron microscope, X-ray diffractometer, ultracentrifuge, spectrophotometer, chromatograph, protein synthesizer, DNA sequencer, etc. These equipments, by their very nature, are concerned with the material, particulate, and molecular aspects of the systems to which they are applied. The bulk of our biological "big data", obtained with these instruments, are thus *physicochemical data*. A scientific theory is supposed to explain observed data. When the data have an overwhelming physicochemical bias, is there any wonder that audacious reductionism claims that a study of life that is not grounded in physicochemical terms cannot be biology? It is completely natural, when the data are "stacked", to have fallen into this trap.

The "scientific method" in modern Western science is a descendent of Baconian empiricism. The tried-and-true routine procedure is to decompose a system under study into simpler parts. These parts must necessarily be (B1) *simple* enough that a complete understanding is possible and (B2) such that the properties of the whole original system may be *reconstructed* from the knowledge of the parts. One often also adds a requisite condition of *universality* that (B3) the *same* analytic procedure may be applied to any system and will in all cases yield a set of parts satisfying (B1) + (B2). In biology, this leads to the scheme

(1)

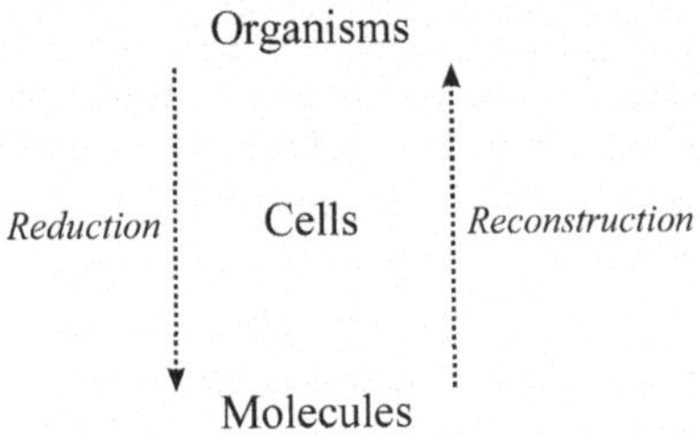

To say that all the properties of the organism must be ultimately resolved into the properties of its constituent molecules constitutes precisely a *reduction* of biology to physics; the postulated universality of the molecular constituents turns *analysis* into *reductionism* (The gross simplification of biological organization into three hierarchical levels in (1) admits, naturally, extrapolations and interpolations; but regardless of the number of levels, the reductionistic scheme of attempting to understand the whole by parts persists.). That this bottom-up strategy, i.e. (B1) +(B2)+(B3), works is reflected in the success of "molecular biology" in the second half of the twentieth century. The analytic parts do get even smaller in chemistry and physics, of course; but I wonder if it will ever come to pass that biological analysis needs to descend further, into, for instance, the realm of elementary particle physics! What if a physical theory of biology falls short even then?

One notes that the algorithmic procedure of analysis into molecules is applicable to *any* material system. Since, to rephrase what I have already stated, a molecule is a molecule is a molecule, reductionism erodes the intuitive distinction one makes between the living and the nonliving. This is not a surprising consequence, since molecular biologists and biochemists almost always experiment with nonliving matter, and in the rare occasions, when they do begin with a living organism, their first step is to kill it. To recapture the essence of the living, the working reductionist would need ad hoc considerations in addition to the inanimate analytic parts. Reductionism fails because the conditions necessary to put Humpty Dumpty together again are often intangible.

Molecular biology and biochemistry are mainstream scholarly pursuits in understanding the chemistry of life, but one must never forget that biology is the science of the living. To this end, one may be pardoned for the quip that the term "molecular biology" is an oxymoron. It is now the twenty-first century; molecular biology has had its run. The pendulum has swung back to approaches that emphasize the behaviour of the whole, that is, *holism* (it is not a dirty word!). For a story of how physiology has returned to centre stage, I would recommend the excellent book *The Music of Life* [Noble 2006] by the systems [sic] biologist Denis Noble, and its sequel *Dance to the Tune of Life* [Noble 2017].

(2)

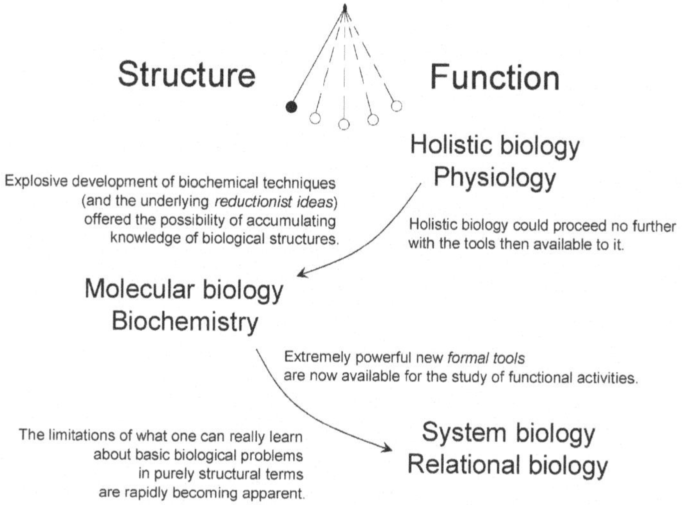

Nicolas Rashevsky (1899–1972) developed the strategy of *relational biology* to capture both the integrative aspects of individual organisms and the unity of the organic world. A motivation is the concept of the *universality* of a set of analytic units into which an organism is to be decomposed. Note that universality is "condition (B3)" of the Baconian method mentioned above. The difference here is the nature of these analytic units: the elements in relational biology are behaviours, and in the reductionist strategy, they are structural subunits. Indeed, the in-kind difference is that of *intangibility* versus tangibility. The idea of universality of behaviour must begin with the acknowledgment of our fundamental intuition that all organisms are in some sense *similar* to one another, and dissimilar to nonliving systems. The basis of biology itself rests on the fact that living systems share common characteristics that are absent (or at least incomplete) in the nonliving. The precise articulation of these characteristics had, however, been elusive; one could not satisfactorily define "life", particularly so if one insisted on attempting to formulate such definitions entirely in structural terms. Rashevsky thus (in the 1950s) concluded that whatever the specific characteristics of organisms might be, they had to involve *relations* rather than structures. An organism is hence an entailment network of its processes; the universal analytic units in the decompositions are mappings, the nature of relational interactions among which being what characterizes a living system. Stated otherwise, integrated homological behaviours of organisms involve (R1) *biological processes* (i.e. *biological functions* that are metabolism, respiration, reproduction, repair, etc.), (R2) *relations* connecting these biological processes, and (R3) *interactions* of these processes with the environment. Properties of the whole *emerge* from the *integration* of (R1), (R2), and (R3). *A fortiori*, knowledge of the components alone tells a limited, partial, and deficient story.

The strategy of relational biology is, in sum, to study biology as organization of relations and is succinctly expressed in our slogan

"*Throw away the matter and keep the underlying organization*".

Robert Rosen (1934–1998) continued the line of succession in relational biology and, among many innovations, instituted the algebraic theory of categories as the natural mathematical metalanguage of the relational theory of life. His work culminated in

The Fundamental Theorem of Relational Biology:
A natural system is an organism if and only if it is closed to efficient causation.

A second motivation for relational biology arises through the employment of *models* for the study of biological processes and their interconnections. Underlying the modelling approach is the concept that *structurally dissimilar systems can nevertheless similarly behave*. Further, *the manifestations of behaviours can also be similar*. Systems with a common model are *analogs* of one another, in which case one can learn about one system by correspondingly studying another. The relational strategy is fundamentally a comparative strategy. Contrariwise, reductionism is intrinsically not comparative, and similar behaviours in different systems can only be individually explained *post hoc*.

I emphasize that the relational strategy is our *approach* to our subject, and the *subject* is *mathematical biology*. While 'biology' is the noun and 'mathematical' is the adjective, it is important to note that both biology and mathematics are indispensable ingredients. Their essential integration is due to the fact that what is common between structurally diverse but behaviourally similar systems can best (and often, only) be formulated in abstract terms, in the form of a mathematical system. Such a mathematical system represents a formal *model*, and each of the natural systems it describes is a *realization*.

$$\textbf{Realization} \underset{\textit{Encoding}}{\overset{\textit{Decoding}}{\rightleftarrows}} \textbf{Model}. \tag{3}$$

Mathematical analysis of the formal models, or *metaphors*, then generates a family of elementary mathematical units, or subsystems, in terms of which the abstract behaviours of the metaphors may be understood. Such an analytic process decodes back into realizations and induces corresponding analytic decompositions in the realizations. Formal subsystems thus play the same role for functional organization that the particulates and molecules play for the reductionists' physicochemical structure. The vital (in every sense of the term) characteristics of the relational approach are (i) the analytic subsystems for a particular behaviour need not (and in general do not) correspond to the structural subunits, and (ii) different behaviours require a decomposition of the same system into different sets of analytic subsystems.

Relational biology proposes that a biological system S be studied through the *reconciliation of alternate descriptions*. One sample pair of alternate descriptions are a 'join of parts' $\vee P$ and a 'whole' W, both representing the same system S, in different modes but on 'the same level'. The relationship (or interactions) between these two alternate, non-equivalent descriptions may be represented as the following modelling relation:

(4)

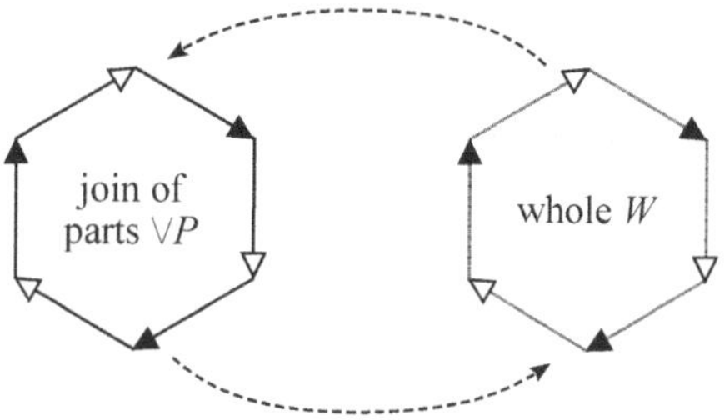

When the system S is a multicellular organism, for example, $\vee P$ is its collection of constituent cells, and W is the whole organism. More generally, when S is a living system, $\vee P$ is its genotype, and W is its phenotype. For all systems S, $W \geq \vee P$. The system S is *simple* when $W = \vee P$, and the system S is *complex* when $W > \vee P$. Neither the join of parts $\vee P$ nor the whole W is sufficient to completely characterize S; the craft of the modelling lies in the interaction and integration of these alternate descriptions. (For an introduction to the $\vee P + W$ relational method, see Louie & Poli [2011]. For a detailed exposition of the partition of the universe of natural systems into simple systems and complex systems, see *ML*: Chapter 9.)

A recurrent reproach to metaphoric methods in biology is that they are "remote from experiment". Even for biologists open to theoretical approaches, they tend to want "biologically realistic (read physicochemical) models". One reason is that 'experiment' is tacitly equivocated to 'physical analysis'. Most experimental data in biology, as I have already mentioned, are collected through instruments designed to *measure* the material, structural (read *tangible*) aspects of a system (biological or otherwise). Such instruments are not designed to *perceive* relations or homologies that are the essence of metaphors. It must be recognized that our perception of such relations is itself a physical act. Construction of relational models is therefore as much an embodiment of physical observation as physical measurements. To narrowly interpret "reality" as the results of structural observations is to shackle oneself in the reductionist's bondage. "Unless I … put my hand in his side, I will not believe".

Another reason for the apparent gap between experimental and theoretical biologists is that structural experiments apply to *specific* systems, while metaphoric approaches apply to *families* of related systems. A bridging resolution will lie in the establishment of relations between the structural (particulate and molecular) analytic subunits of the experimenters and the relational analytic subsystems of the theorists. We (in the Rashevsky–Rosen school of relational biology) have sometimes been asked by experimenters why we do not propose explicit experiments for them to perform and subject our approaches to verification at their hands. We do not do so because it is precisely physicochemical particulars that are abstracted away in the process of generating relational models. To rephrase the vital characteristics of the relational approach: (i′) there is no kind of one-to-one relationship between relational, functional organizations and the structures that realize them; (ii′) a functional organization cuts across physical structures, and a physical structure is simultaneously involved in a variety of functional activities. So an (M,R)-system is not realized by identifying its components and maps in a 'concrete' biological example. To tackle the biological realization problem of (M,R)-systems, one ought not to be seeking physicochemical implementations of what the relations *are*, but, rather, one ought to be seeking interpretations of what the relations *do*. The basic questions of biology, in our view, are not empirical, quantitative questions at all, but, rather, conceptual, qualitative ones. 'Conceptual' and 'qualitative' are not indicators of facileness, but, rather, of *genericity*.

As always, when one attempts to do theory, one is confronted with the trivia: Is it testable, and if so, how? People have been brought up with the prejudice that a scientific theory that is not testable has no merits. It is often considered part of the theorist's job to make theory verifiable, in effect to construct some kind of experimental protocol for the sole purpose of falsification. As long as 'experimental test' exclusively takes on the conventional sense that prescribes 'to falsify some kind of specific physicochemical operation on individual systems', there is in principle no way that the relational descriptions could in fact be 'tested'. A scientific truth cannot be proven; it can only be falsified. A scientific 'truth' can only mean 'there is not sufficient evidence that it is false'. A mathematical truth is absolute: (provable) truths may be proven, and a proven truth is forever true (within the logic system in which the proof has been given). (The cautionary parentheses are due to Gödel.)

A relational description of an organism is as valid, as realistic, a description as any conventional physicochemical one. But it is a description pertaining to a class of physically diverse (though functionally equivalent) systems. A well-constructed model creates a reality of its own: there is no model-independent test of reality. There are many more kinds of experiments than just the physicochemical ones. Conceptual experiments are common in psychology and sociology, for example. Biology has a lot to learn from social sciences.

An act of observation is a quintessential act of abstraction. The measurement of a single quantity of a natural system is indeed the greatest kind of abstraction that can be made of that system: that of analytic modelling by a single real number. The development of theoretical science is thus a synthesis: an attempt to combine observations in such a way that our view of systems becomes *less* abstract than it could be if we were restricted to observation alone. There is no antagonism between 'theory' and 'experiment'. That unfortunate mirage is an artefact of the antagonism between 'theorists' and 'experimenters'.

Relational biology is '*decoding* from formal system to realization'. Experimenters need something to verify, couched in terms of some specific observation, or physicochemical experiment, that they can perform; they need the '*encoding* from natural system to model'. It is through the *synergy of encoding and decoding* that 'theory' and 'experiment', whence 'theorists' and 'experimenters' reconcile their models. If we can agree about our models, we can agree about everything else.

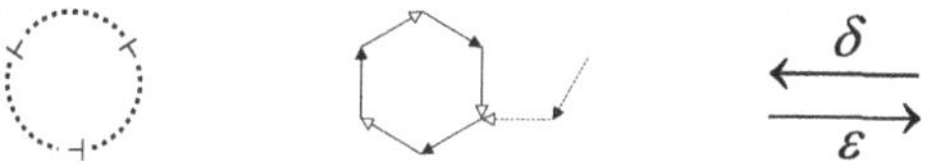

The cast and crew of mathematical and biological characters in *ML*, *More Than Life Itself*, include partially ordered sets, lattices, simulations, models, Aristotle's four causes, graphs, categories, simple and complex systems, anticipatory systems, and metabolism-repair (M,R)-systems. The eminent system theorist George J. Klir reviewed *ML* in *International Journal of General Systems* [Klir 2010]. The journal *Axiomathes* (the theme of which is 'Where Science Meets Philosophy'), edited by Roberto Poli, dedicated an issue [volume 21 number 3, September 2011; Poli 2011] to discussing the nuances of *ML*. Entitled 'Essays on *More Than Life Itself* ', the special topical issue comprises four essays commenting on *ML* and my responses [Louie 2011] to these comments.

RL, *The Reflection of Life*, expands the cast and crew to employ set-valued mappings, adjacency matrices, random graphs, and interacting entailment networks. The anticipation guru Mihai Nadin reviewed *RL*, also in *International Journal of General Systems* [Nadin 2015].

If the theme of Opus I, *ML*, is *one* (M,R)-system, then the theme of Opus II, *RL*, is *two* interacting (M,R)-systems. One, two, many. In this Opus III, the theme is *plurality*. The cast and crew herein iteratively contain those of Opera I & II, plus many notions from category theory, notably equivalence and adjunction.

The encoding and decoding in the modelling relation (e.g. as they appear explicitly and implicitly in diagrams (3) and (4) above) are very examples of *functors*, generalized mapping that has an object component and a process component. This monograph, *Intangible Life: Functorial Connections in Relational Biology* (henceforth denoted by the canonical symbol *IL*), deals with a multitude of *functors* and their *connections*, in their roles in the *relational* approach to *biology*. Alternate descriptions $\vee P$ and W are connected through the power set and the graph, functors that transcend *levels* of reality. Encoding and decoding are inverse operations. In the sequential expansions of the meaning of *invertibility*, their functorial connections recruit many more, including the free and the forgetful.

Thank you, readers, for coming along as companions on my journey. It has been fun. Vouchsafe me a word whether you feel my $\langle ML, RL, IL \rangle$ trilogical cycle has contributed to a comprehensive—or at least comprehensible, if somewhat intangible—inquiry into the nature, origin, and fabrication of life. I may be reached at connect@ahlouie.com.

A. H. Louie
28 May, 2017

Contents

Part III Dimissio: From Invertibility to Adjunction

Prolegomenon
Category Theory for the Aspiring Relational Biologist

> We take the viewpoint that the study of natural systems is precisely the specification of the observables belonging to such a system, and a characterization of the manner in which they are linked. Indeed, for us *observables are the fundamental units of natural systems* …
>
> — Robert Rosen (1985)
> *Anticipatory Systems: Philosophical, Mathematical, and Methodological Foundations*
> 2.1 The Concept of a Natural System

Category

> Category theory asks of every type of Mathematical object: "What are the morphisms?"; it suggests that these morphisms should be described at the same time as the objects.
>
> — Saunders Mac Lane (1997)
> *Category Theory for the Working Mathematician*
> § I. Notes

Robert Rosen entered Nicolas Rashevsky's Committee on Mathematical Biology at the University of Chicago in the autumn of 1957. Engaged in his work on relational biology, Rosen quickly discovered the (M,R)-systems, and developed some of their extraordinary properties. A happy happenstance was when Rosen connected this relational theory of biological systems to the algebraic theory of categories (founded by Samuel Eilenberg and Saunders Mac Lane in 1945), thus equipping himself with a ready-made mathematical tool. Indeed, Rosen's first published scientific paper was on his (M,R)-systems [Rosen 1958a], and his

A. H. Louie, *Intangible Life*, Anticipation Science 2,
DOI 10.1007/978-3-319-65409-6_1

second paper was on 'The representation of biological systems from the standpoint of the theory of categories' [Rosen 1958b].

The confluence of ideas, as can be seen from the above quotes of Rosen and Mac Lane, is that in describing systems, be it natural or formal, the material and efficient causes must be characterized together. The pairs of causes are variously manifested as

(*a*) objects and morphisms;
(*b*) states and observables;
(*c*) structure and function;
(*d*) material and functional entailments;
(*e*) sequential and hierarchical composites;
(*f*) metabolism and repair;
etc.

A category comprises of two collections: i. objects, and ii. morphisms. One may define a category in which the collection of morphisms is partitioned into hom-sets:

0.1 Definition A (*ML*: A.1; *RL*: 6.7) A *category* **C** consists of

i. A collection of *objects*.
ii. For each pair of **C**-objects A, B, a set

(1) $$\mathbf{C}(A,B),$$

the *hom-set* of *morphisms* from A to B. [If $f \in \mathbf{C}(A,B)$, one also writes $f : A \to B$ and $A \xrightarrow{f} B$. Often for simplicity, or when the category **C** need not be emphasized, the hom-set $\mathbf{C}(A,B)$ may be denoted by $H(A,B)$.]

iii. For any three objects A, B, C, a *mapping*

(2) $$\circ : \mathbf{C}(A,B) \times \mathbf{C}(B,C) \to \mathbf{C}(A,C)$$

taking $f : A \to B$ and $g : B \to C$ to its *composite* $g \circ f : A \to C$.

iv. For each object A, there exists a morphism

(3) $$1_A \in \mathbf{C}(A,A),$$

called the *identity morphism* on A.

These entities satisfy the following three axioms:

(c1) *Uniqueness*:

(4) $$\mathbf{C}(A,B) \cap \mathbf{C}(C,D) = \varnothing$$

unless $A = C$ and $B = D$. [Thus each morphism $f : A \to B$ uniquely determines its *domain* $A = \mathrm{dom}(f)$ and *codomain* $B = \mathrm{cod}(f)$: different hom-sets are mutually exclusive.]

(*c*2) *Associativity*: If $f : A \to B,\ g : B \to C,\ h : C \to D$, so that both $h \circ (g \circ f)$ and $(h \circ g) \circ f$ are defined, then

(5) $$h \circ (g \circ f) = (h \circ g) \circ f.$$

(*c*3) *Identity*: For each object A, the identity morphism on A, $1_A : A \to A$, has the property that for any $f : A \to B$ and $g : C \to A$,

(6) $$f \circ 1_A = f \quad \text{and} \quad 1_A \circ g = g$$

[which leads demonstrably to the uniqueness of 1_A in $\mathbf{C}(A,A)$].

Alternatively, one may define a category in terms of arrows, equipping the collection of morphisms with a pair of mappings that assign to each morphism a domain and a codomain:

0.2 Definition B (*RL*: 6.8) A *category* **C** consists of

i′. A set $\mathfrak{O}\mathbf{C}$ of *objects*.

ii′. A set $\mathfrak{A}\mathbf{C}$ of *arrows* (*morphisms*), equipped with two mappings dom and cod:

(7) $$\begin{cases} \mathrm{dom} : & \mathfrak{A}\mathbf{C} \to \mathfrak{O}\mathbf{C} \\ \mathrm{cod} : & \mathfrak{A}\mathbf{C} \to \mathfrak{O}\mathbf{C} \end{cases}.$$

iii′. A (*sequential*) *composition* mapping

(8) $$\circ : \mathfrak{A}\mathbf{C} \times_{\mathfrak{O}\mathbf{C}} \mathfrak{A}\mathbf{C} \to \mathfrak{A}\mathbf{C}$$

(where the domain

(9) $$\mathfrak{A}\mathbf{C} \times_{\mathfrak{O}\mathbf{C}} \mathfrak{A}\mathbf{C} = \{(f,g) \in \mathfrak{A}\mathbf{C} \times \mathfrak{A}\mathbf{C} : \mathrm{dom}(g) = \mathrm{cod}(f)\}$$

is a proper subset of $\mathfrak{A}\mathbf{C} \times \mathfrak{A}\mathbf{C}$, called the '*product over* $\mathfrak{O}\mathbf{C}$', and an ordered pair $(f, g) \in \mathfrak{A}\mathbf{C} \times_{\mathfrak{O}\mathbf{C}} \mathfrak{A}\mathbf{C}$ is called a '*composable pair of morphisms*'), taking (f, g) to its *composite* $g \circ f$, such that

(10) $$\mathrm{dom}(g \circ f) = \mathrm{dom}(f) \quad \text{and} \quad \mathrm{cod}(g \circ f) = \mathrm{cod}(g).$$

iv′. A mapping

(11) $$\mathrm{id} : \mathfrak{O}\mathbf{C} \to \mathfrak{A}\mathbf{C}$$

that sends a C-object A to the *identity morphism* $\mathrm{id}(A) = 1_A$ on A, such that

(12) $$\mathrm{dom}(1_A) = \mathrm{cod}(1_A) = A.$$

These entities satisfy the following two axioms:

(c2′) *Associativity*: If $(f, g) \in \mathfrak{A}\mathbf{C} \times_{\mathfrak{O}\mathbf{C}} \mathfrak{A}\mathbf{C}$ and $(g, h) \in \mathfrak{A}\mathbf{C} \times_{\mathfrak{O}\mathbf{C}} \mathfrak{A}\mathbf{C}$, so that both $h \circ (g \circ f)$ and $(h \circ g) \circ f$ are defined, then

(13) $$h \circ (g \circ f) = (h \circ g) \circ f.$$

(c3′) *Identity*: For any $f : A \to B$, $g : C \to A$, one has

(14) $$f \circ 1_A = f, \quad 1_A \circ g = g.$$

The hom-set $\mathbf{C}(A, B)$ is the inverse image of the pair of $\mathbf{C}$-objects A, B under the mapping $\mathrm{dom} \times \mathrm{cod} : \mathfrak{A}\mathbf{C} \to \mathfrak{O}\mathbf{C} \times \mathfrak{O}\mathbf{C}$:

(15) $$\begin{aligned} \mathbf{C}(A, B) &= (\mathrm{dom} \times \mathrm{cod})^{-1}((A, B)) \\ &= \mathrm{dom}^{-1}(A) \cap \mathrm{cod}^{-1}(B) \\ &= \{f \in \mathfrak{A}\mathbf{C} : \mathrm{dom}(f) = A,\ \mathrm{cod}(f) = B\}. \end{aligned}$$

And the collection $\mathfrak{A}\mathbf{C}$ of morphisms is the disjoint union

(16) $$\mathfrak{A}\mathbf{C} = \bigcup_{A, B \in \mathfrak{O}\mathbf{C}} \mathbf{C}(A, B).$$

For other nuances [e.g., why there is no Axiom ($c1'$)] of the interplay between these two definitions of category and their consequences, see *RL*: 6.7–6.11.

0.3 Associativity Axioms ($c2$) and ($c2'$) imply parentheses are unnecessary in sequential compositions, and the composite in (5) and (13) may simply be denoted

$$h \circ g \circ f : A \to D. \tag{17}$$

The equivalence is illustrated in the *commutative diagram*

(18)

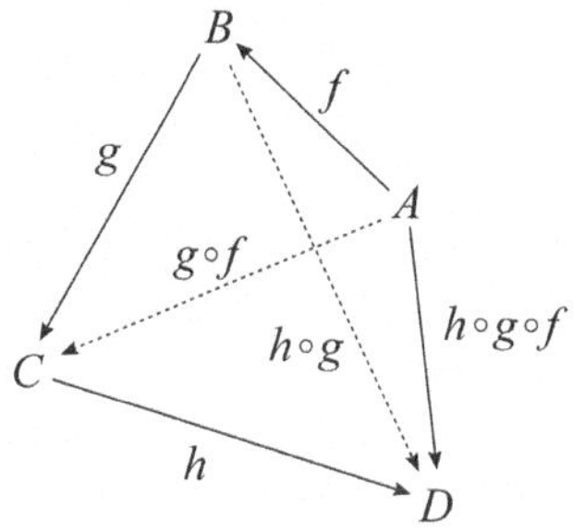

which is a graphical representation that the four paths

$$\begin{cases} A \xrightarrow{g \circ f} C \xrightarrow{h} D \\ A \xrightarrow{f} B \xrightarrow{h \circ g} D \\ A \xrightarrow{h \circ g \circ f} D \\ A \xrightarrow{f} B \xrightarrow{g} C \xrightarrow{h} D \end{cases} \tag{19}$$

trace the same morphism in $\mathbf{C}(A, D)$.

0.4 Categorical Examples

Example i. Note that the only morphisms that are required to exist are the identities on the objects. When there are no objects, there are no identity morphisms. So trivially there is the *empty* category $\varnothing$, with no objects and no morphisms. The next trivial category $\mathbf{C}$ contains exactly one object A and the single identity morphism 1_A, i.e., $\mathfrak{O}\mathbf{C} = \{A\}$ and $\mathfrak{A}\mathbf{C} = \mathbf{C}(A, A) = \{1_A\}$.

Example ii. The correspondence $A \leftrightarrow 1_A$ is a bijection between $\mathfrak{O}\mathbf{C}$ and the subset of identity morphisms in $\mathfrak{A}\mathbf{C}$. The simplest nonempty category is one in which every morphism is an identity, where $\mathbf{C}(A,B) = \varnothing$ when $A \neq B$, and $\mathbf{C}(A,A) = \{ 1_A \}$. Such a category $\mathbf{C}$ is called *discrete*. Every set X is the set of objects of a discrete category $\mathbf{C}$, with $\mathfrak{O}\mathbf{C} = X$ and $\mathfrak{A}\mathbf{C} = \{ 1_x : x \in X \}$.

Example iii. A *monoid* is an algebraic structure with an associative binary operation and an identity element. For any category $\mathbf{C}$ and any $\mathbf{C}$-object X, the hom-set $\mathbf{C}(X,X)$ is a monoid (with the binary operation the composition of $\mathbf{C}$-morphisms, and the identity 1_X). Indeed, a monoid M is a category $\mathbf{C}$ with one object, such that $\mathfrak{O}\mathbf{C} = \{M\}$ and $\mathfrak{A}\mathbf{C} = M$.

Example iv. A *preorder* $\leq$ is a reflexive and transitive relation on a set X ($\leq \subset X \times X$; cf. *ML*:1.10). A *preordered set* $\langle X, \leq \rangle$ may be considered as a category, in which the objects are elements of X, and a hom-set $\mathbf{C}(x,y)$ for $x,y \in X$ has either a single element or is empty, according to whether $x \leq y$ or not. The identity $1_x \in \mathbf{C}(x,x)$ is reflexivity $x \leq x$, and the composition $\circ : \mathbf{C}(x,y) \times \mathbf{C}(y,z) \to \mathbf{C}(x,z)$ is transitivity that $x \leq y$ and $y \leq z$ imply $x \leq z$. In sum, $\mathfrak{O}\mathbf{C} = X$ and $\mathfrak{A}\mathbf{C} = \leq$. A preordered set is a category $\mathbf{C}$ in which the mapping dom $\times$ cod : $\mathfrak{A}\mathbf{C} \to \mathfrak{O}\mathbf{C} \times \mathfrak{O}\mathbf{C}$ ($f \mapsto (\mathrm{dom}(f), \mathrm{cod}(f))$ as in Definition 0.2ii′ above) is injective. This implies that each hom-set $\mathbf{C}(x,y)$ contains at most one morphism; a category with this property is called *thin*. Thus categories with larger hom-sets may be considered to 'generalize' preorders: each morphism defines a distinct preorder relation.

Preorders include *partial orders* (preorders with the additional antisymmetry axiom that $x \leq y$ and $y \leq x$ imply $x = y$; cf. *ML*: 1.20) and *total* (or *linear*) *orders* (partial orders such that, for all $x,y \in X$, either $x \leq y$ or $y \leq x$; cf. *ML*: 1.32). For a *partially ordered set* (*poset*) considered as a category $\mathbf{C}$, the antisymmetry means that if both $\mathbf{C}(x,y)$ and $\mathbf{C}(y,x)$ are nonempty then $x = y$; a category with this property is called *skeletal*. For a *totally ordered set* (*toset*) considered as a category $\mathbf{C}$, the total order means that for all $x,y \in X$, either $\mathbf{C}(x,y)$ or $\mathbf{C}(y,x)$ is nonempty (but if both are nonempty then $x = y$).

Example v. The category **Set** has its collection of objects the set of all sets (in a suitably naive universe of small sets), and its morphisms are mappings from one small set to another. Let me explain *en passant* the phrase 'a suitably naive universe of small sets'. One assumes the existence of a suitable universe U of sets, and then describe a set as a *small set* if it is a member of U. 'Suitable' simply means U has to be big enough for one's purpose, so that the set-theoretic constructions, used in contexts that occur naturally in mathematics, will exist, but U is not too big as to give rise to paradoxical contradictions. This is set theory from the "naive" point of view, and is the common approach of most mathematicians (other than, of course, those in mathematical logic and the foundations of mathematics). In other words, one (aspiring relational biologist included) acknowledges these paradoxes, and moves on.

In a category **C**, the **C**-objects are not necessary sets and the **C**-morphisms are not necessary mappings. But the category **Set** involves itself in an essential way in every category. This is because $\mathfrak{O}$**C** and $\mathfrak{A}$**C** themselves are (for most purposes) sets. Composition and identities are defined by mappings (from a set to a set; Definitions 0.2 iii′ & iv′). Above all, for each pair of **C**-objects A and B, the hom-*set* of **C**-morphisms $\mathbf{C}(A, B)$ is a set.

Example vi. The category **Mon** has its collection of objects the set of all monoids, and its morphisms are monoid homomorphisms from one monoid to another (that preserve the structure of the associative binary operation and the identity). The category **Pos** has as its collection of objects the set of all posets, and its morphisms are order-preserving (isotone) maps from one poset to another (cf. *ML*: 1.23).

Note the difference between the 'single-object-as-a-category' and the 'category of all objects-with-structure and structure-preserving morphisms' considered in the examples above. Contrast a single-set-as-a-category (i.e. a discrete category) with the category **Set** of all sets and mappings. Likewise, contrast a single-monoid-as-a-category (i.e., a single-object category) with **Mon**, and a skeletal category with **Pos**.

0.5 Isomorphism (*ML*: A.5) A morphism $f : A \to B$ is an *isomorphism* if there exists an *inverse morphism* $g : B \to A$ such that $g \circ f = 1_A$ and $f \circ g = 1_B$. If such an inverse morphism exists, it is unique, and is denoted by f^{-1}.

An isomorphism with the same object A as domain and codomain is an *automorphism* on A. If there exists an isomorphism from A to B then A is *isomorphic* to B, and this relation is denoted by

(20) $$A \cong B.$$

Isomorphic objects are considered abstractly (and often identified as) the same, and most constructions of category theory are 'unique up to isomorphism' (in the sense that two similarly constructed objects are isomorphic, if not necessarily identical). The isomorphism relation $\cong$ is an equivalence relation on the collection $\mathfrak{O}$**C** of objects in a category. So instead of "A is isomorphic to B" one may simply say "A and B are isomorphic" by symmetry.

In the category **Set** (of sets and single-valued mappings), isomorphism is the concept of *equipotence* (*RL*: 0.5, et seq. on cardinality); two sets are **Set**-isomorphic precisely when there exists a bijection between them.

0.6 Subcategory (*ML*: A.7) Given categories **C** and **D**, one says that **C** is a *subcategory* of **D** if each **C**-object is a **D**-object, each **C**-morphism is a **D**-morphism, and compositions of morphisms are the same in the two categories. Thus $\mathfrak{O}\mathbf{C} \subset \mathfrak{O}\mathbf{D}$, and for any two **C**-objects A and B, $\mathbf{C}(A,B) \subset \mathbf{D}(A,B)$ (whence *a fortiori* $\mathfrak{A}\mathbf{C} \subset \mathfrak{A}\mathbf{D}$).

More formally, a *subcategory* **C** of a category **D** is given by

i. a subset $X \subset \mathfrak{O}\mathbf{D}$ of **D**-objects, and
ii. a subset $\Phi \subset \mathfrak{A}\mathbf{D}$ of **D**-morphisms,

such that

(s1) for every $A \in X$, the identity morphism $1_A \in \Phi$;
(s2) for every morphism $f : A \to B$ in Φ, both the domain A and the codomain B are in X; and
(s3) for every pair of morphisms f and g in Φ, the composite $g \circ f$ is in Φ whenever it is defined.

These conditions ensure that **C** is a category in its own right: the collection of **C**-objects is $\mathfrak{O}\mathbf{C} = X$, the collection of **C**-morphisms is $\mathfrak{A}\mathbf{C} = \Phi$, and the identities and composition are as in **D**.

If $\mathbf{C}(A,B) = \mathbf{D}(A,B)$ holds for all **C**-objects A and B, **C** is a *full subcategory* of **D**. A full subcategory is one that includes *all* **D**-morphisms between objects of **C**. For any collection $X \subset \mathfrak{O}\mathbf{D}$ of **D**-objects, there is a unique full subcategory **C** of **D** with $X = \mathfrak{O}\mathbf{C}$.

Functor

> **functor** (noun): from Latin *functus*, past participle of the verb *fungi* "to perform" (not the same as the *fungi* meaning yeasts and molds). The Indo-European root is *bheug-* "to enjoy". … [*-or* "a male person or thing that does the indicated action".] A functor is a mapping from one category into another that is compatible with it; the Latin word means literally "performer".
>
> — Steven Schwartzman (1994)
> *The Words of Mathematics: An Etymological Dictionary of Mathematical Terms Used in English*

A functor is a morphism of categories, a mapping from one category to another that preserves the structures and processes therein. A category is defined by the roles of its four cast members: objects, morphisms, composition, identities. A functor, in its performance, must therefore suitably relate these four roles.

0.7 Definition A (*ML*: A.10) Let **C** and **D** be categories. A (*covariant*) *functor* F from **C** to **D**, $F:\mathbf{C}\to\mathbf{D}$, consists of a pair of mappings $\langle F:\mathfrak{O}\mathbf{C}\to\mathfrak{O}\mathbf{D},\ F:\mathfrak{A}\mathbf{C}\to\mathfrak{A}\mathbf{D}\rangle$ on the categorical 'components' of objects and morphisms, called respectively the *object mapping* and the *arrow mapping*, that assigns

i. to each **C**-object A a **D**-object FA,

$$F:A\mapsto FA, \tag{21}$$

and

ii. to each **C**-morphism $f:A\to B$ a **D**-morphism $Ff:FA\to FB$

$$F:[f:A\to B]\mapsto[Ff:FA\to FB]. \tag{22}$$

The object mapping $F:\mathfrak{O}\mathbf{C}\to\mathfrak{O}\mathbf{D}$ and the arrow mapping $F:\mathfrak{A}\mathbf{C}\to\mathfrak{A}\mathbf{D}$ are related in such a way that

(*f*1) if $g\circ f$ is defined in **C**, then $Fg\circ Ff$ is defined in **D**, with

$$F(g\circ f)=Fg\circ Ff; \tag{23}$$

and

(*f*2) for each **C**-object A,

$$F1_A=1_{FA}. \tag{24}$$

Category theory is a formal image of the modelling process itself. It is, indeed, the *general* theory of modelling relations, and not just some *specific* way of making models of one thing in another. It thus generates mathematical counterparts of epistemologies, entirely within the formal realm. One may think of the functor $F:\mathbf{C}\to\mathbf{D}$ as providing, for the category **C**, a *model* $F(\mathbf{C})$ in another category **D**, of all the **C**-objects and **C**-morphisms.

The object mapping $F:\mathfrak{O}\mathbf{C}\to\mathfrak{O}\mathbf{D}$ maps material causes in **C** to material causes in **D**; the arrow mapping $F:\mathfrak{A}\mathbf{C}\to\mathfrak{A}\mathbf{D}$ maps efficient causes in **C** to efficient causes in **D**. The pairwise functorial connection thus extends to the various manifestations; whence $F:\mathfrak{O}\mathbf{C}\to\mathfrak{O}\mathbf{D}$ maps structures to structures, material entailment to material entailment, and $F:\mathfrak{A}\mathbf{C}\to\mathfrak{A}\mathbf{D}$ maps functions to functions, repair to repair, etc.

0.8 Injection and Surjection The functor $F:\mathbf{C}\to\mathbf{D}$ is *injective on objects* if the object mapping $F:\mathfrak{O}\mathbf{C}\to\mathfrak{O}\mathbf{D}$ is injective, and is *surjective on objects* if $F:\mathfrak{O}\mathbf{C}\to\mathfrak{O}\mathbf{D}$ is surjective. Similarly, $F:\mathbf{C}\to\mathbf{D}$ is *injective* (respectively, *surjective*) *on arrows* (or *on morphisms*) if the arrow mapping $F : \mathfrak{A}\mathbf{C} \to \mathfrak{A}\mathbf{D}$ is injective (respectively, surjective).

In set theory, equality of sets is formulated as the Axiom of Extension (*ML* : 0.2): *Two sets are equal if and only if they have the same elements.* (Hence, a priori, two elements of a set are either equal or not.) The object mapping $F : \mathfrak{O}\mathbf{C} \to \mathfrak{O}\mathbf{D}$ is surjective if, by definition, for each **D**-object X there exists a **C**-object A such that $X = FA$. When the requirement of **D**-object-equality is relaxed to **D**-isomorphism, one generalizes the property of surjectivity on objects: a functor $F : \mathbf{C} \to \mathbf{D}$ is *essentially surjective on objects* if for each **D**-object X there exists a **C**-object A such that $X \cong FA$. And of course, if a functor is surjective on objects then it is essentially surjective on objects. 'Essential injectivity on objects', on the other hand, has finer nuances, and its various degrees shall, indeed, be important contributing characteristics towards invertibility.

Property (*f*2), that a functor $F : \mathbf{C} \to \mathbf{D}$ maps an identity morphism in **C** to an identity morphism in **D**, implies that the arrow mapping $F : \mathfrak{A}\mathbf{C} \to \mathfrak{A}\mathbf{D}$ entails the object mapping $F : \mathfrak{O}\mathbf{C} \to \mathfrak{O}\mathbf{D}$. This is because, when the arrow mapping $F : \mathfrak{A}\mathbf{C} \to \mathfrak{A}\mathbf{D}$ takes the value $F1_A = 1_X \in \mathbf{D}(X,X)$ at the **C**-morphism $1_A \in \mathbf{C}(A,A)$, with the correspondence $X \leftrightarrow 1_X$ one may uniquely define the object mapping $F : \mathfrak{O}\mathbf{C} \to \mathfrak{O}\mathbf{D}$ to take the value $FA = X$ at the **C**-object A.

A functor, just like a category, may alternatively be defined in terms of arrows (without the redundant postulate i′ for the object mapping):

0.9 Definition B A (*covariant*) *functor* F from category **C** to category **D**, $F : \mathbf{C} \to \mathbf{D}$, is

ii′. a mapping $F : \mathfrak{A}\mathbf{C} \to \mathfrak{A}\mathbf{D}$ of arrows that sends $f \in \mathfrak{A}\mathbf{C}$ to $Ff \in \mathfrak{A}\mathbf{D}$,

(25) $$F : f \mapsto Ff,$$

carrying

(*f*1′) each composable pair of **C**-morphisms $(f,g) \in \mathfrak{A}\mathbf{C} \times_{\mathfrak{O}\mathbf{C}} \mathfrak{A}\mathbf{C}$ to a composable pair of **D**-morphisms $(Ff,Fg) \in \mathfrak{A}\mathbf{D} \times_{\mathfrak{O}\mathbf{D}} \mathfrak{A}\mathbf{D}$, with

(26) $$F(g \circ f) = Fg \circ Ff;$$

and

(*f*2′) each identity morphism in $\mathfrak{A}\mathbf{C}$ to an identity morphism in $\mathfrak{A}\mathbf{D}$.

Often, for the sake of clarity, however, one explicitly specifies the action of a functor on both objects and arrows.

0.10 Functorial Representation A functor $F : \mathbf{C} \to \mathbf{D}$ may be succinctly represented in

(27) $$F : \begin{cases} A \mapsto FA & (A \in \mathfrak{O}\mathbf{C}) \\ [f : A \to B] \mapsto [Ff : FA \to FB] & (f \in \mathfrak{A}\mathbf{C}) \end{cases},$$

the two lines denoting respectively the object mapping $F : \mathfrak{O}\mathbf{C} \to \mathfrak{O}\mathbf{D}$ and the arrow mapping $F : \mathfrak{A}\mathbf{C} \to \mathfrak{A}\mathbf{D}$.

As denoted in (27), the general representation does not, of course, provide additional information about F. Its use lies in the specific forms that the final causes FA and $Ff : FA \to FB$ would take for specific functors under study. Then representation (27) provides a concise summary of the actions of the functor F.

0.11 Contravariant Functor Besides the covariant functors there is a dual kind of functors that reverses the direction of the processes and the order of composition. A *contravariant functor* F from **C** to **D** assigns

i. to each **C**-object A a **D**-object FA,

(28) $$F : A \mapsto FA,$$

and

ii ^{op}. to each **C**-morphism $f : A \to B$ a **D**-morphism $Ff : FB \to FA$

(29) $$F : [f : A \to B] \mapsto [Ff : FB \to FA],$$

such that

($f1^{\mathrm{op}}$) if $g \circ f$ is defined in **C**, then $Ff \circ Fg$ is defined in **D**, and

(30) $$F(g \circ f) = Ff \circ Fg.$$

and

($f2$) for each **C**-object A,

(31) $$F1_A = 1_{FA}.$$

Its succinct representation is

(32) $$F : \begin{cases} A \mapsto FA & (A \in \mathfrak{O}\mathbf{C}) \\ [f : A \to B] \mapsto [Ff : FB \to FA] & (f \in \mathfrak{A}\mathbf{C}) \end{cases}.$$

0.12 Hom-Functors (*ML*: A.13) For any category **C** and a **C**-object A, the *covariant hom-functor* $h^A = \mathbf{C}(A, \bullet)$ from **C** to **Set** assigns to each **C**-object Y the set $h^A Y = \mathbf{C}(A, Y)$, and to a **C**-morphism $k : Y \to Y'$ the mapping $h^A k : \mathbf{C}(A, Y) \to \mathbf{C}(A, Y')$ defined by

(33) $$h^A k : f \mapsto k \circ f \quad \text{for} \quad f : A \to Y;$$

i.e. via the diagram

(34)

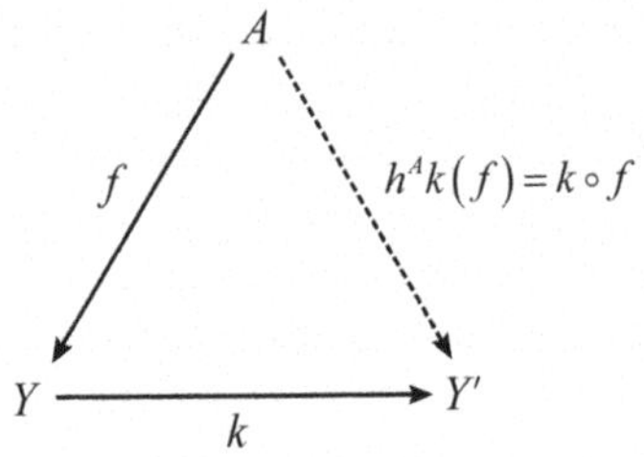

Note the action of $h^A k$ may be described as 'composition with k-on-the-left'.

Dually, for a category **C** and a **C**-object B, the *contravariant hom-functor* $h_B = \mathbf{C}(\bullet, B)$ assigns to each **C**-object X the set $h_B X = \mathbf{C}(X, B)$, and to a **C**-morphism $g : X \to X'$ the mapping $h_B g : \mathbf{C}(X', B) \to \mathbf{C}(X, B)$ defined by

(35) $$h_B g(f) = f \circ g \quad \text{for} \quad f : X' \to B;$$

i.e. via the diagram

(36)

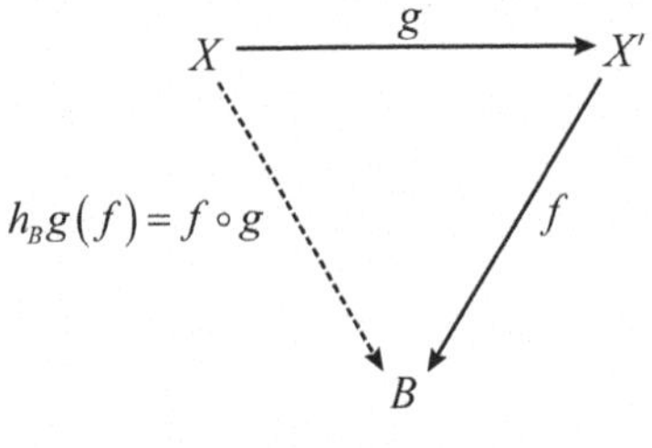

Note the action of $h_B g$ may be described as 'composition with g-on-the-right'.

0.13 The Category Cat (*ML*: A.15) The idea of category applied to categories and functors themselves yields the category **Cat**, with objects all categories (i.e. all *small* categories in a suitably naïve universe) and morphisms all functors between them.

Functors can be composed—given functors $F : \mathbf{C} \to \mathbf{D}$ and $G : \mathbf{D} \to \mathbf{E}$, the maps $A \mapsto G(FA)$ and $f \mapsto G(Ff)$ on **C**-objects A and **C**-morphisms f define a functor $G \circ F : \mathbf{C} \to \mathbf{E}$. This composition is associative, since it is associative componentwise on objects and morphisms. For each category **C** there is an *identity functor* $I_{\mathbf{C}} : \mathbf{C} \to \mathbf{C}$, defined in the natural way as the identity map componentwise, sending each **C**-object to itself and each **C**-morphism to itself.

An *isomorphism* $F : \mathbf{C} \to \mathbf{D}$ of categories is a functor that is a bijection both on objects and on morphisms. This is equivalent to the existence of an 'inverse functor' $F^{-1} : \mathbf{D} \to \mathbf{C}$.

0.14 Faithful and Full Functors (*ML*: A.16) For each pair of **C**-objects A and B, the functor $F : \mathbf{C} \to \mathbf{D}$ assigns to each **C**-morphism $f \in \mathbf{C}(A, B)$ a **D**-morphism $Ff \in \mathbf{D}(FA, FB)$, and so defines a (single-valued) mapping

(37) $$F_{A,B} : \mathbf{C}(A, B) \to \mathbf{D}(FA, FB)$$

with $F_{A,B}(f) = Ff$. The functor may alternatively be considered as the collection of these doubly-indexed mappings:

(38) $$F = \{ F_{A,B} : A, B \in \mathfrak{O}\mathbf{C} \}.$$

The functor F is *faithful* when each $F_{A,B}$ is injective, and *full* when each $F_{A,B}$ is surjective.

Faithfulness and fullness are functorial conditions on the arrow mapping $F : \mathfrak{A}\mathbf{C} \to \mathfrak{A}\mathbf{D}$, and each by itself does not impose limitations on the object mapping $F : \mathfrak{O}\mathbf{C} \to \mathfrak{O}\mathbf{D}$. So a faithful functor need not be injective on objects: two **C**-objects may map to the same **D**-object. Likewise, a full functor need not be surjective on objects: there may be **D**-objects not of the form FA for some $A \in \mathfrak{O}\mathbf{C}$.

Injectivity on arrows is a stronger condition than faithfulness: if $F : \mathbf{C} \to \mathbf{D}$ is injective on arrows then it is faithful. But the converse implication is not true: a faithful functor need not be injective on arrows. The collection of **C**-hom-sets $\{ \mathbf{C}(A, B) : A, B \in \mathfrak{O}\mathbf{C} \}$ forms a partition of $\mathfrak{A}\mathbf{C}$ (cf. (16) above), and faithfulness only requires that the *restriction* of the arrow mapping to each block $\mathbf{C}(A, B)$, $F_{A,B} = F|_{\mathbf{C}(A,B)} : \mathbf{C}(A, B) \to \mathbf{D}(FA, FB)$, be injective, whereas injectivity on arrows requires $F : \mathfrak{A}\mathbf{C} \to \mathfrak{A}\mathbf{D}$ to be injective on the whole domain $\mathfrak{A}\mathbf{C}$. A functor $F : \mathbf{C} \to \mathbf{D}$ that is faithful may still map two **C**-morphisms with different domains or codomains, (therefore belonging to different **C**-hom-sets) to the same **D**-morphism. Injectivity on arrows also implies injectivity on objects. This is because, if $F : \mathbf{C} \to \mathbf{D}$ is injective on arrows, then in particular, for $A, B \in \mathfrak{O}\mathbf{C}$ and $A \neq B$, F must map the distinct $1_A, 1_B \in \mathfrak{A}\mathbf{C}$ to distinct $1_{FA}, 1_{FB} \in \mathfrak{A}\mathbf{D}$, whence $FA \neq FB$ in $\mathfrak{O}\mathbf{D}$.

Similarly, surjectivity on arrows implies surjectivity on objects: if $F : \mathbf{C} \to \mathbf{D}$ is surjective on arrows, for each $X \in \mathfrak{O}\mathbf{D}$ there is an $f \in \mathfrak{A}\mathbf{C}$ that gets mapped by $F : \mathfrak{A}\mathbf{C} \to \mathfrak{A}\mathbf{D}$ to $1_X \in \mathbf{D}(X,X) \subset \mathfrak{A}\mathbf{D}$, thence both $\mathrm{dom}(f), \mathrm{cod}(f) \in \mathfrak{O}\mathbf{C}$ (which need not coincide) are mapped by $F : \mathfrak{O}\mathbf{C} \to \mathfrak{O}\mathbf{D}$ to X. Further, if a functor is surjective on arrows then it is full, hence contrapositively a functor that is not full cannot be surjective on arrows. Conversely, a full functor $F : \mathbf{C} \to \mathbf{D}$ need not be surjective on arrows: **D**-morphisms between **D**-objects that are not of the form FA for some $A \in \mathfrak{O}\mathbf{C}$ cannot come from **C**-morphisms.

Even if $F : \mathbf{C} \to \mathbf{D}$ is both faithful and full, whence each mapping $F_{A,B} : \mathbf{C}(A,B) \to \mathbf{D}(FA,FB)$ is bijective, the collection $\{F_{A,B} : A, B \in \mathfrak{O}\mathbf{C}\}$ of **Set**-isomorphisms is still not sufficient to ensure that F is an isomorphism in the category **Cat**. As explicated above, the range $F(\mathbf{C})$ is not necessarily isomorphic to either **C** or **D**. A faithful and full functor is, however, necessarily injective on objects *up to isomorphism*. When $F : \mathbf{C} \to \mathbf{D}$ is a faithful and full functor, one may readily verify, using the definition of isomorphism and the premise that all mapping $F_{A,B} : \mathbf{C}(A,B) \to \mathbf{D}(FA,FB)$ are then bijections, that $FA \cong FB$ implies $A \cong B$. This defines one version of 'essentially injective on objects'.

0.15 Inclusion Functor (*ML*: A.12(v)) If **C** is a subcategory of **D**, there is a functor that takes objects and morphisms to themselves; i.e., both the object mapping and the arrow mapping are the corresponding inclusion maps. This is the *inclusion functor* (of **C** in **D**), denoted $i : \mathbf{C} \to \mathbf{D}$.

The inclusion functor $i : \mathbf{C} \to \mathbf{D}$ is injective on objects, injective on arrows, and faithful. It is full if and only if **C** is a full subcategory of **D**.

0.16 Concrete Category and Forgetful Functor A *concrete category* **C** is a category equipped with a faithful functor $F : \mathbf{C} \to \mathbf{Set}$. The faithfulness of F allows the (one-to-one) identification of a **C**-morphism $f \in \mathfrak{A}\mathbf{C}$ with the mapping $Ff \in \mathfrak{A}\mathbf{Set}$. A concrete category may be described as a category **C** in which each **C**-object A comes equipped with an 'underlying set' FA, each **C**-morphism $f \in \mathbf{C}(A,B)$ is an actual mapping $Ff : FA \to FB$, and the composition of **C**-morphisms is a composition of mappings. Stated otherwise, the faithful functor $F : \mathbf{C} \to \mathbf{Set}$ allows the consideration of **C**-objects as *sets with additional* *structure*, and of **C**-morphisms as *structure-preserving mappings*. The functor $F : \mathbf{C} \to \mathbf{Set}$ then, in essence, 'forgets' the additional structure of the objects and hence the structure-preserving aspect of the mappings; it is therefore called the *forgetful functor*.

Many important categories have interpretations as concrete categories; for example, the category **Grp** of groups and homomorphisms, the category **Vct** of vector spaces and linear transformations, and the category **Top** of topological spaces and continuous mappings (*ML*: A.6).

The requirement for a concrete category **C** is that the functor $F : \mathbf{C} \to \mathbf{Set}$ be faithful, but not necessarily injective on arrows. This means that F must take different morphisms in $\mathbf{C}(A, B)$ to different mappings in $\mathbf{Set}(FA, FB)$, but it may take different **C**-objects to the same set, since injectivity on objects is not a requirement (say $A, B \in \mathfrak{O}\mathbf{C}$, $A \neq B$, but the sets $FA = FB$). If this occurs, it will also take corresponding **C**-morphisms in $\mathbf{C}(A, Y)$ and $\mathbf{C}(B, Y)$, for example, to the same mapping in $\mathbf{Set}(FA, FY) = \mathbf{Set}(FB, FY)$.

0.17 Membership and Element-Tracing In a concrete category **C**, one may speak of 'membership' $a \in A$ for a **C**-object $A \in \mathfrak{O}\mathbf{C}$, and 'element chase' $f : a \mapsto b = f(a)$ associated with a **C**-morphism $f : A \to B$ where $f \in \mathfrak{A}\mathbf{C}$. (For the element-trace notation $f : a \mapsto f(a)$ see *ML*: 1.5 and *RL*: 1.7; I shall also re-introduce it in *IL*: Chapter 2.)

When $F : \mathbf{C} \to \mathbf{D}$ is a functor between concrete categories, the object mapping $F : \mathfrak{O}\mathbf{C} \to \mathfrak{O}\mathbf{D}$ at $A \in \mathfrak{O}\mathbf{C}$, $F : A \mapsto FA$, hierarchically entails the element mapping $F_A : A \to FA$. The action of the arrow mapping $F : \mathfrak{A}\mathbf{C} \to \mathfrak{A}\mathbf{D}$, taking $f : A \to B$ to $Ff : FA \to FB$, may then be represented in the commutative diagram

(39)

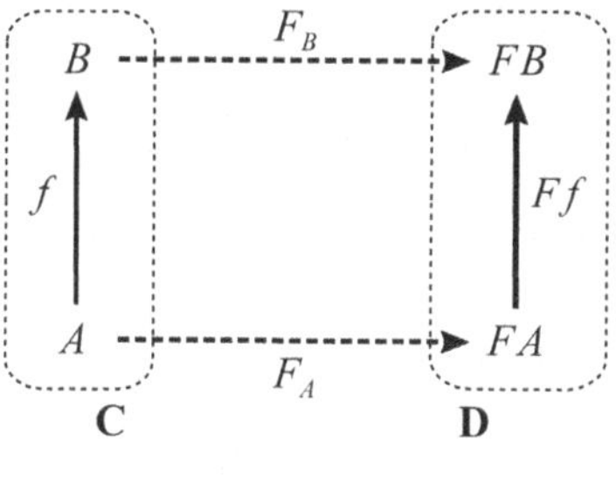

which declares the equality of two sequential compositions

(40)
$$F_B \circ f = Ff \circ F_A : A \to FB.$$

The compositional equality entails, for $a \in A$ and the traces of the paths

(41) $$\begin{cases} a \mapsto f(a) \mapsto F_B(f(a)) \\ a \mapsto F_A(a) \mapsto Ff(F_A(a)) \end{cases},$$

elemental equality of the final causes, resulting in

(42) $$F_B(f(a)) = Ff(F_A(a)) \in FB.$$

The corresponding element-trace diagram is

(43)

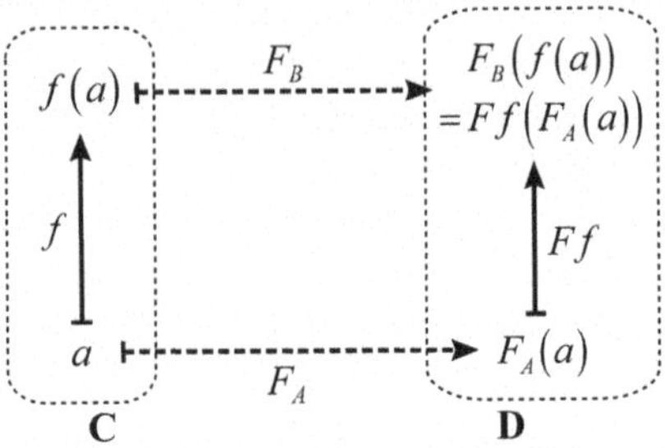

In terms of the solid-headed and hollow-headed arrows of a *relational diagram in graph-theoretic form* (*ML*: 5.4–5.11; *RL*: E.6 & 3.1; and, in anticipation, *IL*: 2.2), the confluence of two sequential compositions (40) is represented thus:

(44)

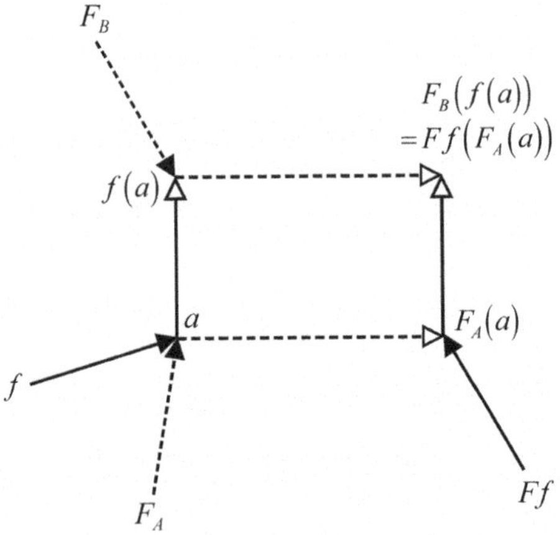

Natural Transformation

> … "category" has been defined in order to be able to define "functor" and "functor" has been defined in order to be able to define "natural transformation".
>
> — Saunders Mac Lane (1997)
> *Category Theory for the Working Mathematician*
> § I.4

A natural transformation is a morphism of functors (*ML*: A.17). This is the vehicle with which one functor models another.

0.18 Definition Suppose

(45)
$$\mathbf{C} \overset{F}{\underset{G}{\rightrightarrows}} \mathbf{D}$$

are two functors between the same two categories. A *natural transformation* τ from F to G, notated

(46)
$$\tau : F \to G,$$

i. assigns to each $\mathbf{C}$-object A a $\mathbf{D}$-morphism $\tau_A \in \mathbf{D}(FA, GA)$,

such that,

($t1$) for each $\mathbf{C}$-morphism $f \in \mathbf{C}(A, B)$, the $\mathbf{D}$-morphisms $Gf \in \mathbf{D}(GA, GB)$, $\tau_A \in \mathbf{D}(FA, GA)$, $\tau_B \in \mathbf{D}(FB, GB)$, and $Ff \in \mathbf{D}(FA, FB)$ commute:

(47)
$$Gf \circ \tau_A = \tau_B \circ Ff.$$

Graphically, this is the commutative diagram

(48)

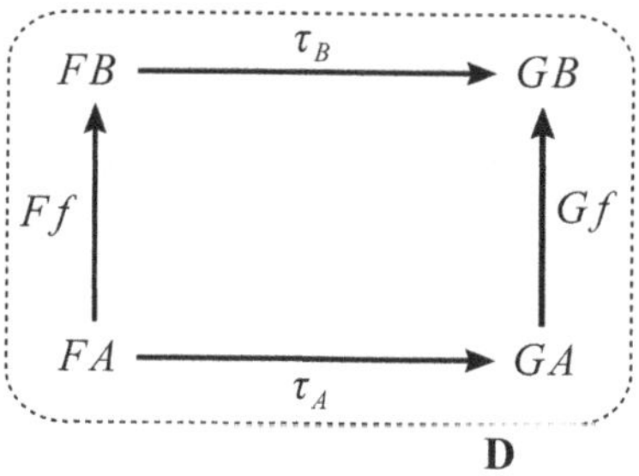

τ_A is called the *component of* τ at A.

Note the antitone decrease in the numbers of requirements as one ascends the hierarchy: a category (Definition 0.1) has four assignments and three properties (c1)–(c3); a functor (Definition 0.7) has two assignments and two properties (f1)–(f2); a natural transformation has one assignment i and one property (t1).

A natural transformation $\tau : F \to G$ may be considered to be determined by the collection of components

(49) $$\{\tau_A \in \mathbf{D}(FA, GA) : A \in \mathbf{OC}\}.$$

$\tau_A \in \mathbf{D}(FA, GA)$ is said to be *natural in A*, in the sense that when the **C**-object A is treated as a variable, the **D**-morphism $\tau_A \in \mathbf{D}(FA, GA)$ is 'defined in the same way for each A'. This is the standard terminology ("informal parlance") of a more proper "$\tau_{(\bullet)} : F(\bullet) \to G(\bullet)$ is natural in its variable".

Since a functor $F : \mathbf{C} \to \mathbf{D}$ gives a picture (or model) in **D** for any collection of objects and morphisms of **C**, one may consider a natural transformation $\tau : F \to G$ to be a translation (alternate description or model) of the picture F to the picture G. For example, picture (18), the commutative diagram of **C**-morphism associativity, has the following translation from F to G:

(50)

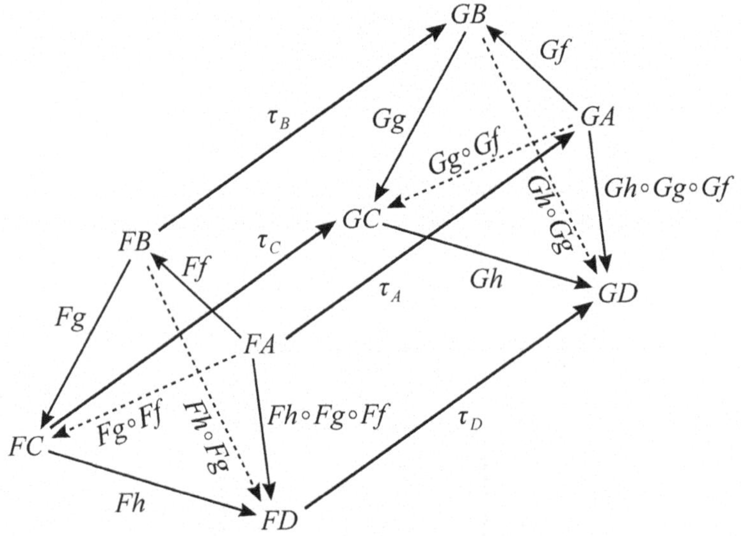

a digraph of **D**-morphisms in which all paths are commutative (i.e., any two directed paths with the same initial and final vertices trace the same morphism).

0.19 Functor Category The *functor category* $\mathbf{D}^{\mathbf{C}}$ has as objects all (covariant) functors from **C** to **D**, and as morphisms natural transformations, and to have composition and identities the 'pointwise' ones (*ML*: A.18).

0.20 Natural Isomorphism A natural transformation $\tau : F \to G$ is a *natural isomorphism*, denoted

(51) $$\tau : F \cong G,$$

if and only if for each **C**-object A, $\tau_A \in \mathbf{D}(FA, GA)$ is an isomorphism in **D**.

Stated otherwise, a natural isomorphism is an isomorphism in the functor category $\mathbf{D}^{\mathbf{C}}$.

0.21 Category of Diagrams If **C** is a trivial category with only a single object A and only the single morphism 1_A in $\mathbf{C}(A,A)$ (Example 0.4i), then the functor category $\mathbf{D}^{\mathbf{C}}$ is a discrete category (Example 0.4ii), consisting of the objects of **D** together with their identity morphisms. That is, $\mathfrak{O}\mathbf{D}^{\mathbf{C}} \cong \mathfrak{O}\mathbf{D}$ and $\mathfrak{A}\mathbf{D}^{\mathbf{C}} \cong \{1_X : X \in \mathfrak{O}\mathbf{D}\} \cong \mathfrak{O}\mathbf{D}$.

Next, let **C** consist of a pair of objects A, B, and suppose that the morphisms in **C** consist only of 1_A, 1_B, and a single morphism $f : A \to B$. Then given any other category **D**, the functor category $\mathbf{D}^{\mathbf{C}}$ may be regarded as consisting of all the morphisms in **D**; i.e., $\mathfrak{O}\mathbf{D}^{\mathbf{C}} \cong \mathfrak{A}\mathbf{D}$.

A graphic interpretation is as follows: the category **C** may be regarded as being specified by the simple diagram

(52) $$A \xrightarrow{f} B.$$

(The identity morphisms correspond to self-loops (*ML*: 6.3) on the objects, and may be omitted.) The functor category $\mathbf{D}^{\mathbf{C}}$ consists of all copies of this diagram in **D**; i.e., all diagrams of the form

(53) $$X \xrightarrow{g} Y,$$

where $X = FA$, $Y = FB$, $g = Ff$ for some covariant functor $F : \mathbf{C} \to \mathbf{D}$. More illustratively, when the category **C** is concrete and $f : A \to B$ is a mapping, the relational diagram in graph-theoretic form of (52) may be drawn as

(54) 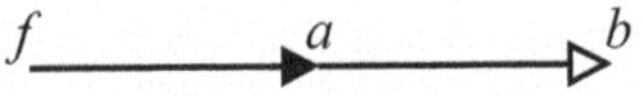

with corresponding relational-diagrammatic representation

(55) 

in **D**. Thus, if the category **C** is regarded as specifying the 'pattern' (54), the functor category $\mathbf{D}^{\mathbf{C}}$ consists of all copies of this pattern which may be formed in **D**.

More generally, *any* diagram of **C**-morphisms (i.e., a network) in a category **C** can be regarded as specifying a subcategory $\mathbf{C}'$ of **C** (with careful inclusion of composites); then the functor category $\mathbf{D}^{\mathbf{C}'}$ (which is a subcategory of $\mathbf{D}^{\mathbf{C}}$) may again be regarded as the collection of copies of this diagram that may be formed from the objects and morphisms of **D**. Hence the larger functor category $\mathbf{D}^{\mathbf{C}}$ contains copies of all **C**-diagrams, and is therefore also called the *category of diagrams* over **C**.

0.22 Binary Operation Let $R : \mathbf{Grp} \to \mathbf{Set}$ be the forgetful functor (Definition 0.16) that sends a group $G \in \mathfrak{O}\mathbf{Grp}$ to its underlying set $RG \in \mathfrak{O}\mathbf{Set}$ and a homomorphism $\varphi \in \mathbf{Grp}(G,H)$ to the mapping $R\,\varphi \in \mathbf{Set}(R\,G, R\,H)$. Let $S : \mathbf{Grp} \to \mathbf{Set}$ be the "Cartesian square functor", defined by

(56) $$S : \begin{cases} G \mapsto RG \times RG & (G \in \mathfrak{O}\mathbf{Grp}) \\ [\varphi : G \to H] \mapsto [R\varphi : RG \times RG \to RH \times RH] & (\varphi \in \mathfrak{A}\mathbf{Grp}) \end{cases},$$

where

(57) $$R\varphi(x,y) = (\varphi \times \varphi)(x,y) = (\varphi x, \varphi y) \quad (x,y \in G).$$

The binary operation $\cdot_G$ of a group $G \in \mathfrak{O}\mathbf{Grp}$ is a mapping

(58) $$\tau_G : RG \times RG \to RG,$$

i.e., $\tau_G \in \mathbf{Set}(RG \times RG, RG) = \mathbf{Set}(SG, RG)$, defined by

(59) $$\tau_G(x,y) = x \cdot_G y \quad (x,y \in G).$$

The family $\tau = \{\tau_G \in \mathbf{Set}(SG, RG) : G \in \mathbf{OGrp}\}$ is a natural transformation from S to R.

(60)

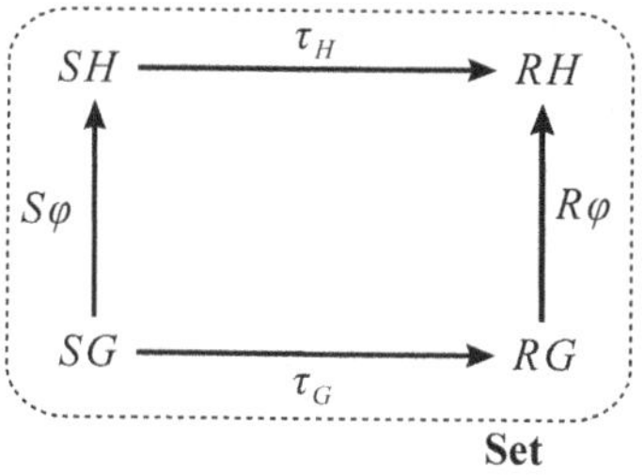

The equality, for $(x,y) \in SG = RG \times RG$, of the traces of the paths

(61)
$$\begin{cases} (x,y) \mapsto (\varphi x, \varphi y) \mapsto \varphi x \cdot_H \varphi y \\ (x,y) \mapsto x \cdot_G y \mapsto \varphi(x \cdot_G y) \end{cases},$$

results in

(62)
$$\varphi x \cdot_H \varphi y = \varphi(x \cdot_G y),$$

which is the defining property of the homomorphism $\varphi : G \to H$. The naturality condition thus simply means that the binary operation in groups can be regarded as a natural transformation.

0.23 Hom-Functors Fix a morphism $f \in \mathbf{C}(A, B)$ in the category $\mathbf{C}$. For each $Y \in \mathbf{OC}$, define the mapping $\tau_Y : \mathbf{C}(B, Y) \to \mathbf{C}(A, Y)$ by

(63)
$$\tau_Y(g) = g \circ f \quad (g \in \mathbf{C}(B, Y)).$$

These mappings are the components of a natural transformation

(64)
$$\tau : h^B \to h^A$$

from the covariant hom-functor $h^B = \mathbf{C}(B, \bullet) : \mathbf{C} \to \mathbf{Set}$ (Section 0.12) to the covariant hom-functor $h^A = \mathbf{C}(A, \bullet) : \mathbf{C} \to \mathbf{Set}$. For a $\mathbf{C}$-morphism $k : Y \to Y'$, the commutative diagram is

(65)

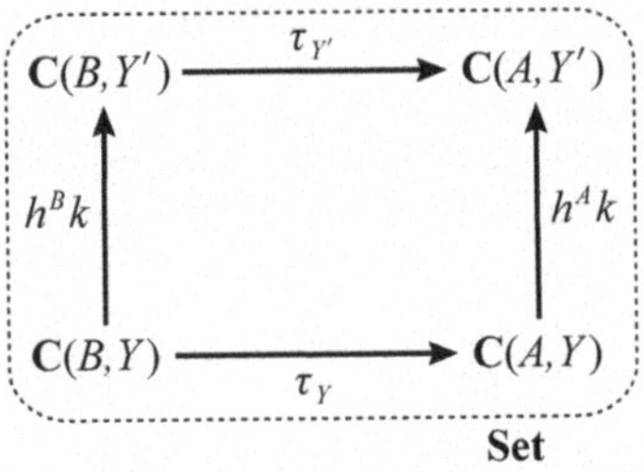

The equality, for $g \in \mathbf{C}(B, Y)$, of the traces of the paths

(66) $$\begin{cases} g \mapsto h^B k(g) = k \circ g \mapsto \tau_{Y'}(k \circ g) = (k \circ g) \circ f \\ g \mapsto \tau_Y(g) = g \circ f \mapsto h^A k(g \circ f) = k \circ (g \circ f) \end{cases},$$

results in

(67) $$(k \circ g) \circ f = k \circ (g \circ f),$$

which is simply the associativity condition of **C**-morphisms.

Recall that the action of $h^{\cdot} k$ may be described as 'composition with k-on-the-left'. The action of $\tau_{\cdot}$ may be described as 'composition with f-on-the-right'. The natural transformation $\tau : h^B \to h^A$ illustrates that associativity holds when, in the 'double composition' $k \circ g \circ f$, the order in which composition with k-on-the-left and composition with f-on-the-right are performed does not affect the result.

Dually, for each $X \in \mathcal{O}\mathbf{C}$, define the mapping $\sigma_X : \mathbf{C}(X, A) \to \mathbf{C}(X, B)$ by

(68) $$\sigma_X(h) = f \circ h \quad (h \in \mathbf{C}(X, A)).$$

These mappings are the components of a natural transformation

(69) $$\sigma : h_A \to h_B$$

from the contravariant hom-functor $h_A = \mathbf{C}(\cdot, A) : \mathbf{C} \to \mathbf{Set}$ to the contravariant hom-functor $h_B = \mathbf{C}(\cdot, B) : \mathbf{C} \to \mathbf{Set}$. The corresponding commutative diagram again illustrates **C**-morphism associativity.

0.24 Evaluation Map For sets X and Y, the set **Set**(X, Y) of all mappings from X to Y is denoted Y^X. The *evaluation* mapping $e : Y^X \times X \to Y$, defined, for $f : X \to Y$ and $x \in X$, by $e(f,x) = f(x)$, may be interpreted as a natural transformation as follows. For a fixed X, the map $Y \mapsto Y^X \times X$ extends to a functor $F : \mathbf{Set} \to \mathbf{Set}$ with, for $g : Y \to Z$, $Fg : Y^X \times X \to Z^X \times X$ defined by $Fg : (f,x) \mapsto (g \circ f, x)$ for $f : X \to Y$ and $x \in X$. Then, for this fixed X, $e : F \to I_{\mathbf{Set}}$ is a natural transformation from the functor F to the identity functor $I_{\mathbf{Set}}$, i.e., the following square commutes for any mapping $g : Y \to Z$:

(70)

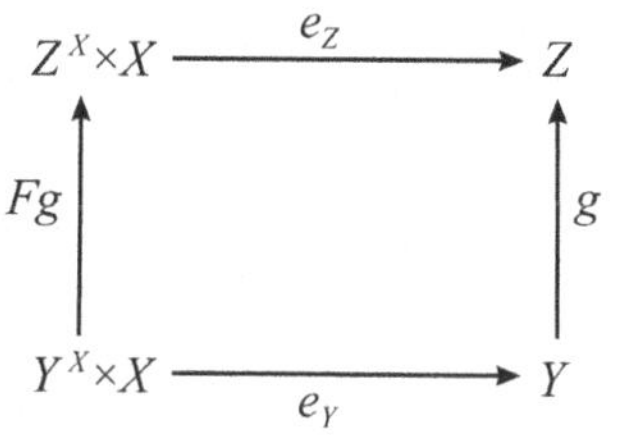

This reduces to the equation $g(e_Y(f,x)) = e_Z(g \circ f, x)$, which says simply that $g(f(x)) = (g \circ f)(x)$.

0.25 Dual Vector Spaces In the category **Vct** of vector spaces over a fixed field K, evaluation takes the following form. Each element $x \in V$ defines an *evaluation mapping* $\hat{x} : V^* \to K$ by $\hat{x}(f) = f(x)$ for every $f \in V^*$. $\hat{x}$ is a linear functional on V^*, hence it is a member of V^{**}, the second dual space of V. The mapping $\alpha_V : V \to V^{**}$ defined by $\alpha_V(x) = \hat{x}$ is an isomorphism (of vector-spaces) when V is finite dimensional. It is called the *natural isomorphism* between V and V^{**}. (Note this linear-algebraic terminology is part of the inspiration for its category-theoretic analogue.) For a linear transformation $T : V \to W$, one has $T^{**} \circ \alpha_X = \alpha_Y \circ T$, i.e., the diagram

(71)

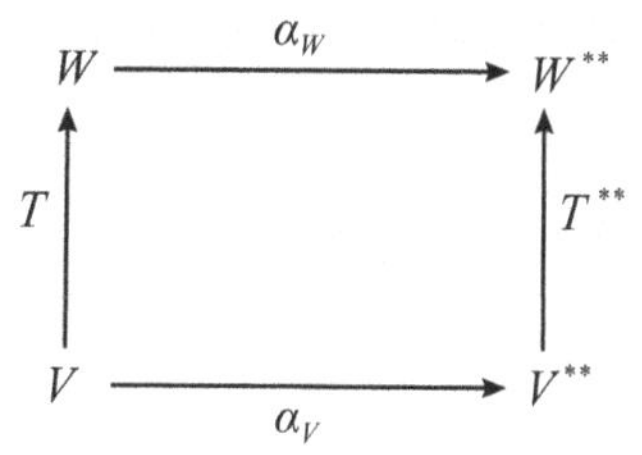

commutes, which says precisely that $\alpha : I_{\mathbf{Vct}} \to (\bullet)^{**}$ is a natural transformation.

0.26 Material and Functional Entailments A mapping of two variables $t : X \times Y \to Z$ may be considered as a mapping $\varphi t : X \to Z^Y$ of one variable (in X), and the values of which are mappings with domain in the second variable (in Y) and codomain in Z:

$$[\varphi t(x)](y) = t(x,y) \quad \text{for} \quad x \in X \quad \text{and} \quad y \in Y. \tag{72}$$

Equality (72) describes φ as a bijection (i.e. an isomorphism in **Set**)

$$\varphi : \mathbf{Set}(X \times Y, Z) \cong \mathbf{Set}\left(X, Z^Y\right) \tag{73}$$

that is natural in X, Y, and Z. The isomorphism (73) may be written as

$$\mathbf{Set}(X \times Y, Z) \cong \mathbf{Set}(X, \mathbf{Set}(Y, Z)) \tag{74}$$

or

$$H(X \times Y, Z) \cong H(X, H(Y, Z)). \tag{75}$$

The last bijection (75), connecting material entailment (metabolism) on the left-hand side with functional entailment (repair) on the right-hand side, is of particular importance in (M,R)-systems. It has wonderful consequences in relational biology, from ontogenesis (*ML*: 13.25) to therapeutics (*RL*: 14.9–14.10). It also leads into the category-theoretic concept of adjunction, and will reappear many times as we proceed in *IL*.

Part I
Potestas
The Power Set Functor

—Doxology of the *Pater Noster*

Ascent

The *power set functor* is the most important functor in relational biology. It plays an indispensable role in the category-theoretic formulation of closure to efficient causation (*RL*: 9.3 & 9.4), the very characterization of life. It is (usually) defined as the covariant functor $\mathrm{P} : \mathbf{Set} \to \mathbf{Set}$ that assigns to a set X its power set $\mathrm{P}X$ and assigns to a mapping $f : X \to Y$ the mapping $\mathrm{P}f : \mathrm{P}X \to \mathrm{P}Y$ that sends each subset $A \subset X$ to its image $(\mathrm{P}f)(A) = f(A) \subset Y$, viz.

$$\mathrm{P} : \begin{cases} X \mapsto \mathrm{P}X & (X \in \mathfrak{O}\mathbf{Set}) \\ [f : x \mapsto f(x)] \mapsto [\mathrm{P}f : A \mapsto f(A)] & (f \in \mathfrak{A}\mathbf{Set}) \end{cases} .$$

The power set functor P is an essential tool in the analysis of impredicative systems through the reconciliation of two alternate descriptions of an impredicative system. Tersely, the entities '$\langle X, f \rangle$' and '$\langle \mathrm{P}X, \mathrm{P}f \rangle$' are alternate descriptions on different 'levels' of the same system 'X'. The mapping $f : X \to Y$ maps on the 'element level' (i.e. parts) while the mapping $\mathrm{P}f : \mathrm{P}X \to \mathrm{P}Y$ maps on the 'set level' (i.e. whole). Thus the power set functor P efficiently ascends hierarchical levels.

On our journey in relational biology, the power set functor $\mathrm{P} : \mathbf{Set} \to \mathbf{Set}$ was first introduced as an example in *ML*: A.12(ii) and explicated in more detail in *RL*: 1.18 *et seq*. I shall presently formulate it alternatively in the category **Rel** of sets and relations.

I would like to share an anecdote. During the algebra session of my PhD comprehensive examination (the other two sessions being analysis and mathematical biology) in the spring of 1980, I was verily grilled by professors on everything I knew about the subject. But my supervisor Robert Rosen asked me exactly one question: 'What are the actions of the power set functor?'

It may therefore be said that thence planted was the intangible seed of the tangible manifestation of a model of the *arbor scientiae* that is this monograph *IL*.

1 Prooemium Relations

> It is idle to talk always of the alternative of reason and faith. Reason is itself a matter of faith. It is an act of faith to assert that our thoughts have any relation to reality at all.
>
> — G.K. Chesterton (1908)
> *Orthodoxy*
> Chapter III. "The Suicide of Thought"

Let me begin with a parody of a few passages from the Prologomenon of *RL*. Expository divergence is, however, imminent …

Sets

1.1 Subset and Superset If A and B are sets and if every element of A is an element of B, then A is a *subset* of B, and B is a *superset* of A, denoted

$$(1) \qquad A \subset B \quad \text{(equivalently,} \quad B \supset A).$$

Note that this symbolism of containment means *either* $A = B$ (which means the sets A and B have the same elements; Axiom of Extension, *ML*: 0.2) *or* A is a *proper subset* of B (which means that B contains at least one element that is not in A). Two sets A and B are equal if and only if $A \subset B$ and $B \subset A$ (*ML*: 0.4).

1.2 Inclusion Map For $A \subset B$, the mapping $i : A \to B$ defined by $i(a) = a$ for all $a \in A$ is called the *inclusion map* (of A in B). If the sets involved need to be emphasized, one may use the notation $i_{A \subset B}$ for the inclusion map. The inclusion map of A in A is called the *identity map* on A, denoted 1_A $(= i_{A \subset A})$.

A. H. Louie, *Intangible Life*, Anticipation Science 2,
DOI 10.1007/978-3-319-65409-6_2

1.3 Equipotence Two sets are *equipotent* (to each other) if there exists a bijective mapping, i.e., a one-to-one correspondence, between them.

1.4 Cardinality A set is *finite* if it is either empty or equipotent to the set $\{0, 1, 2, \ldots, n-1\}$ for a natural number n; otherwise it is *infinite*. An infinite set that is equipotent to the set $\mathbb{N}$ of all natural numbers is called *countably infinite*; otherwise the infinite set is *uncountable*. The term *countable* means either finite or countably infinite.

With the formal definition $0 = \varnothing$ and $n = \{0, 1, 2, \ldots, n-1\}$, a finite set is 'equipotent to a whole number'. Each finite set X is equipotent to a *unique* whole number $|X| = n \in \mathbb{N}_0$, the 'number of elements of X'. In short, a finite set is a set consisting of a finite number of elements.

The property that each finite set is equipotent to a unique whole number may be extended to infinite sets. The generalized 'number of elements' of a set is called its *cardinality*, and formally one has the

1.5 Property Every set is equipotent to a unique *cardinal number*.

The usual partial order $\leq$ of whole numbers may be extended to all cardinal numbers. One uses the same notation $|X| = n$ for the cardinality of the set X, where n may be an 'infinite cardinal' in addition to a whole number. Infinite cardinal numbers are usually denoted by the first letter $\aleph$ (*aleph*) of the Hebrew alphabet, ordered by a nonnegative integer subscript. When $|X| = n$, one may simply say 'X has cardinal number n' or 'X has cardinality n'.

1.6 Theorem

i. *Every set has a cardinal number.*
ii. *Two sets A and B are equipotent if and only if they have the same cardinal number, i.e., iff $|A| = |B|$.*
iii. $|A| \leq |B|$ *if and only if A is equipotent to a subset of B* (*which includes the special case when $A \subset B$*).
iv. $|A| < |B|$ *if and only if A is equipotent to a subset of B but B is not equipotent to a subset of A.*

1.7 Corollaries

i. *Every finite set has a unique number of elements.*
ii. *Two finite sets are equipotent if and only if they have the same number of elements.*
iii. *If a set is finite, then every one of its subsets is finite.*
iv. *If a finite set X has n elements and a subset $A \subset X$ has k elements, then $k \leq n$; further, $k = n$ iff $A = X$.*
v. *If a set is finite, then it is not equipotent to any of its proper subsets.*

Property v, that a finite set is not equipotent to any of its proper subsets, in fact characterizes finite sets. The inverse thus characterizes infinite sets; stated formally:

1.8 Theorem

i. *A set is infinite if and only if it is equipotent to a proper subset of itself.*
ii. *A set is finite if and only if it is not equipotent to any proper subset of itself.*

Sets from Sets

1.9 Complements The *relative complement* of a set A in a set B is the set of elements in B but not in A:

(2) $$B \sim A = \{x \in B : x \notin A\}.$$

Note that this definition does not require that $A \subset B$, and one has $B \sim A = B \sim (A \cap B) \subset B$. If $A \subset B$ and B is finite (whence A also), then $|B \sim A| = |B| - |A|$. When B is the 'universal set' U (of some appropriate universe under study, e.g. the set of all natural systems **N** or the 'largest set' in some field of sets), the set $U \sim A$ is denoted A^c, i.e.

(3) $$A^c = \{x \in U : x \notin A\},$$

and is called simply the *complement* of the set A. An element of U is either a member of A, or not a member of A, but not both. That is, $A \cup A^c = U$, and $A \cap A^c = \varnothing$.

1.10 De Morgan's Laws Union and intersection interchange under complementation: for sets $\{A_i\}_{i \in I}$ and B,

(4) $$\begin{aligned} B \sim \left(\bigcup_{i \in I} A_i \right) &= \bigcap_{i \in I} (B \sim A_i) \\ B \sim \left(\bigcap_{i \in I} A_i \right) &= \bigcup_{i \in I} (B \sim A_i). \end{aligned}$$

1.11 Inclusion–Exclusion Principle The familiar counting equality for finite sets,

(5) $$|A \cup B| = |A| + |B| - |A \cap B|,$$

implies, in particular, the inequality $|A \cup B| \leq |A| + |B|$, with $|A \cup B| = |A| + |B|$ iff $|A \cap B| = |\varnothing| = 0$ (i.e., iff sets A and B are *disjoint*). The results generalize for finite sets $A_1, A_2, \ldots, A_n$ to

(6) $$\begin{aligned} |A_1 \cup A_2 \cup \cdots \cup A_n| = {} & |A_1| + |A_2| + \cdots + |A_n| \\ & - |A_1 \cap A_2| - |A_1 \cap A_3| - \cdots - |A_{n-1} \cap A_n| \\ & + |A_1 \cap A_2 \cap A_3| + |A_1 \cap A_2 \cap A_4| \\ & \qquad + \cdots + |A_{n-2} \cap A_{n-1} \cap A_n| \\ & \vdots \\ & + (-1)^{n-1} |A_1 \cap A_2 \cap \cdots \cap A_n|, \end{aligned}$$

which may be succinctly written as

(7) $$\left| \bigcup_{i=1}^{n} A_i \right| = \sum_{k=1}^{n} (-1)^{k-1} \left(\sum_{1 \leq i_1 < \cdots < i_k \leq n} |A_{i_1} \cap A_{i_2} \cap \cdots \cap A_{i_k}| \right).$$

Further, $\left| \bigcap_{i=1}^{n} A_i \right| \leq \sum_{i=1}^{n} |A_i|$, with equality iff the sets $A_1, A_2, \ldots, A_n$ are pairwise disjoint.

1.12 Power Set If X is a set, the *power set* $\mathrm{P}X$ of X is the family of all subsets of X.

The inclusion relation $\subset$ is a *partial order* on the power set $\mathrm{P}X$; i.e., $\langle \mathrm{P}X, \subset \rangle$ is a *poset* (*ML*: 1.22). The least element of $\langle \mathrm{P}X, \subset \rangle$ is $\varnothing$, and the greatest element of $\langle \mathrm{P}X, \subset \rangle$ is X (*ML*: 1.28). Note that even when $X = \varnothing$, $\varnothing \in \mathrm{P}X$ (indeed, $\mathrm{P}X = \{\varnothing\}$) so $\mathrm{P}X \neq \varnothing$. $\langle \mathrm{P}X, \cup, \cap \rangle$ is a complete, complemented *lattice* (*ML*: 2.1, 2.12, 3.12). $\langle \mathrm{P}X, \cup, \cap, ^c \rangle$ is a Boolean algebra (*ML*: 3.19), called the *power set algebra* of X. A *field of sets* is a subalgebra of a power set algebra. The power set algebra is, indeed, the 'universal' Boolean algebra, in the sense that every Boolean algebra is isomorphic to a field of sets (Stone Representation Theorem, *ML*: 3.20).

1.13 Characteristic Mapping A subset A of X may be identified with its *characteristic mapping*, a mapping χ_A from X to $2 = \{0, 1\}$ defined by

(8) $$\chi_A(x) = \begin{cases} 0 & \text{if } x \notin A \\ 1 & \text{if } x \in A \end{cases}.$$

When X is a finite set with n members, there are 2^n different mappings $\chi : X \to 2$, because for each element $x \in X$ there are precisely two choices for the value $\chi(x)$, either 0 or 1. If one defines $A = \chi^{-1}(1) \subset X$, then $\chi = \chi_A$.

1.14 Cardinality of the Power Set Thus if $|X| = n$, then $|\mathrm{P}X| = 2^n$, and the equality may be extended to all cardinal numbers n, finite and infinite. This gives an alternate notation of the power set $\mathrm{P}X$ as 2^X. One may succinctly write

(9) $$|\mathrm{P}X| = |2^X| = 2^{|X|}.$$

This is consistent even if $X = \varnothing$, when $|X| = 0$ and $|\mathrm{P}X| = 2^0 = 1$. Cantor's Theorem (*RL*: 0.8) states that, for all sets X, $|X| < 2^{|X|}$.

The equivalent notation $\mathrm{P}X = 2^X$ expressing the power set as a 'power' is, of course, the origin of its name.

> **power** (noun): from Old French *poeir*, from Vulgar Latin *potere*, a variant of Classical Latin *posse* "to be able". The Indo-European root is *poti-* "powerful; lord". If you are able to do many things, you are powerful. A powerful person typically has a large number of possessions (a word derived from *posse*) and a large amount of money. In algebra, when even a relatively small number like 2 is multiplied by itself a number of times the result gets large very quickly; metaphorically speaking, the result is powerful. ... If the term *power* is used precisely, it refers to the result of multiplying a number by itself a certain number of times. Consider $2^3 = 8$, which says that the 3rd power of 2 is 8. The power is 8. In less precise usage, however, 3 is identified as the power, when it is actually the exponent.
>
> — Steven Schwartzman (1994)
> *The Words of Mathematics: An Etymological Dictionary of Mathematical Terms Used in English*

1.15 Product Given two sets X and Y, one denotes by $X \times Y$ the set of all *ordered pairs* of the form (x, y) where $x \in X$ and $y \in Y$. The set $X \times Y$ is called the *product* (or *Cartesian product*) of the sets X and Y. If either X or Y is empty, then $X \times Y = \varnothing$.

For all sets X and Y, the cardinality of the product set is the product of the cardinalities of the components:

(10) $$|X \times Y| = |X||Y|.$$

1.16 Projections The mappings

(11) $$\pi_1 : X \times Y \to X \quad \text{and} \quad \pi_2 : X \times Y \to Y,$$

defined, for $x \in X$ and $y \in Y$, by

(12) $$\pi_1(x, y) = x \quad \text{and} \quad \pi_2(x, y) = y,$$

are the *canonical projections* (of the product $X \times Y$ onto its components; cf. *ML*: A.22).

For $A \subset X$, the set $\pi_1^{-1}(A)$ of the *inverse image* of A is the subset of $X \times Y$ containing all ordered pairs (x, y) that are sent by π_1 into A:

(13) $$\pi_1^{-1}(A) = \{(x, y) \in X \times Y : \pi_1(x, y) = x \in A\} = A \times Y.$$

Similarly, for $B \subset Y$,

(14) $$\pi_2^{-1}(B) = \{(x, y) \in X \times Y : \pi_2(x, y) = y \in B\} = X \times B.$$

The product set $A \times B \subset X \times Y$ may be identified with the set $\pi_1^{-1}(A) \cap \pi_2^{-1}(B)$ of intersection of inverse images, since

(15) $$\pi_1^{-1}(A) \cap \pi_2^{-1}(B) = (A \times Y) \cap (X \times B) = A \times B.$$

Relations

1.17 Definition A A *relation* R is an ordered triple (X, Y, Γ) where X and Y are sets and Γ is a subset of the Cartesian product $X \times Y$. The sets X and Y are respectively called the *domain* and *codomain* of the relation, and $\Gamma \subset X \times Y$ is called its *graph*.

One may indicate the dependence of X, Y, and Γ on R with the notations $X = \text{dom}(R)$, $Y = \text{cod}(R)$, and $\Gamma(R)$.

According to the formal Definition 1.17, a relation uniquely determines its domain and codomain, so two relations with identical graphs but different domains or different codomains are considered different. [This is, indeed, the category-theoretic requirement that a morphism uniquely entails its domain and codomain; Definitions 0.1, 0.2, and cf. *ML*: A.1; *RL*: 6.7 et seq.] Consider the simple example $\Gamma = \{(2,\text{A}),(1,\text{C}),(2,\text{B})\}$. The relations $R_1 = (\{1,2,3,4,5\}, \{\text{A, B, C, D, E, F}\},\Gamma)$, $R_2 = (\mathbb{N}, \text{alphanumeric characters}, \Gamma)$, $R_3 = (\mathbb{Z}, \text{Latin alphabet}, \Gamma)$, and $R_4 = (\mathbb{R}, \{\text{A, B, C}\}, \Gamma)$ are all distinct.

A relation is often identified with its graph (hence the minor equivocation $R = \Gamma(R)$), so one also has the (more common but less rigorous)

1.18 Definition B A *relation* is a set R of ordered pairs; i.e. $R \subset X \times Y$ for some sets X and Y.

Equivalently, a relation R is an element of the power set $\mathrm{P}(X \times Y)$, i.e., $R \in \mathrm{P}(X \times Y)$. With domain X and codomain Y, the relation R is *from* X *to* Y. The collection of *all* relations from X to Y is thus the power set $\mathrm{P}(X \times Y)$, and, in view of (9) and (10) above, the cardinality of this collection is

$$|\mathrm{P}(X \times Y)| = 2^{|X \times Y|} = 2^{|X||Y|}. \tag{16}$$

If $(x,y) \in R$ (or more precisely $(x,y) \in \Gamma(R)$), then one may say that x *is R-related to* y (or simply x *is related to* y when the involved relation R is understood).

There is a chirality inherent in $(x,y) \in R \subset X \times Y$. When $X \neq Y$, the asymmetry between a relation from X to Y and a relation from Y to X are apparent. But even when $R \subset X \times X$ (whence $\text{dom}(R) = \text{cod}(R) = X$ and one says R is a *relation on* X), $(x,y) \in R$ and $(y,x) \in R$ (for $x, y \in X$) are independent statements. (See *ML*: 1.9 et seq. for an exposition of the epistemological consequences of relations on X.) To emphasize the chirality inherent in $(x,y) \in R$, one may also say that x is a *left R-relative* (*left relative*) of y, and that y is a *right R-relative* (*right relative*) of x.

1.19 External and Internal Entailments Note that even in the formulation 1.18, a relation still has to uniquely determine its domain and codomain, although these two sets cannot be induced from the components of the ordered pairs that are members of $\Gamma(R) = R$. In the example in 1.17, when $\Gamma(R) = \{(2, \mathrm{A}), (1, \mathrm{C}), (2, \mathrm{B})\}$, from $\Gamma(R)$ itself one may only conclude that $X = \mathrm{dom}(R)$ must be a superset of $\{1, 2\}$ and $Y = \mathrm{cod}(R)$ must be a superset of $\{\mathrm{A}, \ \mathrm{B}, \ \mathrm{C}\}$; no more information is forthcoming *internally* from $\Gamma(R) = R$.

Thus the domain X and the codomain Y of a relation R have to be *externally* supplied; they are objects extraneous to the graph Γ of R. It is therefore more satisfactory (and more accurate) to apply the term 'relation' to the ordered triple $(X, Y, \Gamma) = R$ rather than $\Gamma = R \subset X \times Y$. But if this sort of thing were systematically done, the mathematical notation would become rather cumbersome. Let me quote Rudin [1986: 1.21] on this issue:

> Most mathematical systems are sets with some class of distinguished subsets or some binary operations or some relations (which are required to have certain properties), and one can list these and then describe the system as an ordered pair, triple, etc., depending on what is needed. For instance, the real line may be described as a quadruple $(\mathbb{R}^1, +, \cdot, <)$, where $+$, $\cdot$, and $<$ satisfy the axioms of a complete archimedean ordered field. But it is a safe bet that very few mathematicians think of the real field as an ordered quadruple.

On this note, I shall henceforth use Definition 1.18 for a relation R, with the tacit assumption that the domain X and codomain Y are known, and use the notation $R \subset X \times Y$ or $R \in \mathrm{P}(X \times Y)$ when these sets need to be emphasized.

1.20 Relation Examples The relation $U = X \times Y \in \mathrm{P}(X \times Y)$ is the *universal relation*, in which every $x \in X$ is related to every $y \in Y$. The relation $\varnothing \in \mathrm{P}(X \times Y)$ is the *empty relation*, in which no $x \in X$ is related to any $y \in Y$. The empty relation $\varnothing \subset X \times Y$ is a relation from X to Y for all sets X and Y, even if either X or Y is empty, in which case $X \times Y = \varnothing$ and $\mathrm{P}(X \times Y) = \{\varnothing\}$. In the partially ordered set $\langle \mathrm{P}(X \times Y), \subset \rangle$, U is the greatest element and $\varnothing$ is the least element (*ML*: 1.28): for all relations $R \in \mathrm{P}(X \times Y)$, $\varnothing \subset R \subset U$.

Let X be a set. *Membership* ("is an element of") is the relation $\in_X \subset X \times \mathrm{P}X$ defined, for $x \in X$ and $A \in \mathrm{P}X$, by

$$(17) \qquad (x, A) \in \in_X \quad \text{iff} \quad x \in A.$$

Inclusion ("is a subset of") is the relation $\subset_X \subset \mathrm{P}X \times \mathrm{P}X$ defined, for $A, B \in \mathrm{P}X$, by

(18) $$(A, B) \in \subset_X \quad \text{iff} \quad A \subset B.$$

$\in_X$ contains no members of the form $(x, \varnothing)$ for any $x \in X$. $\subset_X$ contains $(\varnothing, B)$ and (A, X) for all $A, B \in \mathrm{P}X$.

One may note that if $|X| = n$, then $|\mathrm{P}X| = 2^n$, whence

(19) $$|X \times \mathrm{P}X| = n2^n \quad \text{and} \quad |\mathrm{P}X \times \mathrm{P}X| = 2^{2n}.$$

For each $i = 0, 1, \ldots, n$, there are $\binom{n}{i}$ subsets A of X with cardinality i. Each of these A s with $|A| = i$ contains i elements $x \in A$, so there are i ordered pairs of the form (x, A). Thus, as the subset $\in_X \subset X \times \mathrm{P}X$, the cardinality of the membership relation is

(20) $$|\in_X| = \sum_{i=0}^{n} \binom{n}{i} i = n2^{n-1}.$$

One therefore sees that the membership relation contains exactly half of the total number of eligible element–subset ordered pairs:

(21) $$\frac{|\in_X|}{|X \times \mathrm{P}X|} = \frac{n2^{n-1}}{n2^n} = \frac{1}{2}.$$

Each of the $\binom{n}{i}$ subsets B of X with $|B| = i$ itself contains $|\mathrm{P}B| = 2^i$ distinct subsets. Thus, as the subset $\subset_X \subset \mathrm{P}X \times \mathrm{P}X$, the cardinality of the inclusion relation is

(22) $$|\subset_X| = \sum_{i=0}^{n} \binom{n}{i} 2^i = 3^n.$$

The fraction of subsets that satisfy the inclusion relation $A \subset B$ among all possible subset pairs $A, B \in \mathrm{P}X$ is thus

(23)
$$\frac{|\subset_X|}{|\mathrm{P}X \times \mathrm{P}X|} = \frac{3^n}{2^{2n}} = \left(\frac{3}{4}\right)^n.$$

1.21 Corange and Range The *corange* of a relation $R \subset X \times Y$ is the subset of its domain containing all those $x \in X$ for which there is at least one $y \in Y$ such that $(x, y) \in R$. The *range* of a relation $R \subset X \times Y$ is the subset of its codomain containing all those $y \in Y$ for which there is at least one $x \in X$ such that $(x, y) \in R$.

Stated otherwise,

(24)
$$\begin{aligned} \operatorname{cor}(R) &= \{x \in X : \exists y \in Y \ (x, y) \in R\} \subset \operatorname{dom}(R) = X \\ \operatorname{ran}(R) &= \{y \in Y : \exists x \in X \ (x, y) \in R\} \subset \operatorname{cod}(R) = Y \end{aligned}.$$

For the simple example $\Gamma = \{(2, \mathrm{A}), (1, \mathrm{C}), (2, \mathrm{B})\}$, $\operatorname{cor}(R) = \{1, 2\}$ and $\operatorname{ran}(R) = \{\mathrm{A}, \ \mathrm{B}, \ \mathrm{C}\}$. But $\{1, 2, 3, 4, 5\} \times \{\mathrm{A}, \mathrm{B}, \mathrm{C}, \mathrm{D}, \mathrm{E}, \mathrm{F}\}$, $\mathbb{N} \times$ alphanumeric characters, $\mathbb{Z} \times$ Latin alphabet, and $\mathbb{R} \times \{\mathrm{A}, \ \mathrm{B}, \ \mathrm{C}\}$ are among an infinitude of valid possibilities for $\operatorname{dom}(R) \times \operatorname{cod}(R)$.

To emphasize the points made previously, let me rephrase the situation as follows. The corange and range are defined in an 'internal' sense; the relation R (material cause) within itself entails the respective sets (final causes). Indeed, $\operatorname{cor}(R)$ and $\operatorname{ran}(R)$ are simply canonically projected images of the set R of ordered pairs onto its first and second components (cf. Definition 1.16 above):

(25)
$$\pi_1(R) = \operatorname{cor}(R) \quad \text{and} \quad \pi_2(R) = \operatorname{ran}(R).$$

In other words, $R \mapsto \operatorname{cor}(R)$ and $R \mapsto \operatorname{ran}(R)$ are well-defined, algorithmic, *material entailments*, information mechanistically caused 'inside' the relation, from an 'intrinsic' perspective. The domain and codomain, on the other hand, are defined only in an 'external' sense; the efficient causes dom and cod must dictate the respective values. These beyond-syntax assignment rules correspond to *functional entailments* $\vdash$ dom and $\vdash$ cod; they are prescriptively caused from 'outside' of the relation, from an 'extrinsic' perspective.

1.22 Analysis and Synthesis With $A = \mathrm{cor}(R) \subset X$ and $B = \mathrm{ran}(R) \subset Y$ in (15), one has

$$\mathrm{cor}(R) \times \mathrm{ran}(R) = \pi_1^{-1}(\pi_1(R)) \cap \pi_2^{-1}(\pi_2(R)). \tag{26}$$

One notes that $R \subset \mathrm{cor}(R) \times \mathrm{ran}(R)$, but in general R is a proper subset of $\mathrm{cor}(R) \times \mathrm{ran}(R)$, so $R \neq \mathrm{cor}(R) \times \mathrm{ran}(R)$. The material entailment $\pi : R \mapsto \{\mathrm{cor}(R), \mathrm{ran}(R)\}$ is *analysis*, while the attempt at reversal $\pi^{-1} : \{\mathrm{cor}(R), \mathrm{ran}(R)\} \mapsto R$ is *synthesis*. The composition $\pi^{-1} \circ \pi$ in (26) illustrates that, in general, from the analytic components (i.e., information on $x \in \mathrm{cor}(R)$ and $y \in \mathrm{ran}(R)$) one may at best synthesize the superset $\mathrm{cor}(R) \times \mathrm{ran}(R) \supset R$; the relational information on $(x, y) \in R$ has been lost so one cannot recover R itself. Thus

$$\text{synthesis} \circ \text{analysis} \not\equiv \text{identity}. \tag{27}$$

Stated otherwise, once the relation R is broken into its parts, its internal relational connections are lost; it is generically not possible to synthesize R back from its analytic pieces—woe to reductionism! I have explicated this non-invertibility in the context of the amphibology of analysis and synthesis in detail in *ML*: 7.43–7.49; the reader is cordially invited to revise therein. I shall have a lot more to say on invertibility (and the lack thereof that is irreversibility) here in *IL*.

Relational Operations

1.23 Converse Relation The *converse* of a relation $R \subset X \times Y$ is the relation $\breve{R} \subset Y \times X$ such that $(y, x) \in \breve{R}$ if and only if $(x, y) \in R$.

Manifestly, one has

$$\begin{aligned} \mathrm{dom}\left(\breve{R}\right) &= \mathrm{cod}(R) = Y, & \mathrm{cod}\left(\breve{R}\right) &= \mathrm{dom}(R) = X, \\ \mathrm{cor}\left(\breve{R}\right) &= \mathrm{ran}(R), & \mathrm{ran}\left(\breve{R}\right) &= \mathrm{cor}(R). \end{aligned} \tag{28}$$

The graph $\Gamma\left(\breve{R}\right)$ of the converse relation $\breve{R}$ is the *transpose* of the graph $\Gamma(R)$ of R:

(29) $$\Gamma\left(\breve{R}\right) = \{(y,x) \in Y \times X : (x,y) \in R\} = [\Gamma(R)]^t.$$

The converse operation is an involution; the converse of the converse of a relation is the relation itself:

(30) $$\breve{\breve{R}} = R.$$

The converse of the membership relation (or simply *converse membership*) is $\ni_X \subset \mathrm{P}X \times X$ defined, for $A \in \mathrm{P}X$ and $x \in X$, by

(31) $$A \ni_X x, \quad \text{i.e., } (A,x) \in \ni_X, \quad \text{iff} \quad x \in X.$$

The converse of the inclusion relation is the relation 'includes' ('is a superset of') $\supset_X \subset \mathrm{P}X \times \mathrm{P}X$ defined, for $A, B \in \mathrm{P}X$, by

(32) $$(A,B) \in \supset_X \quad \text{iff} \quad A \supset B.$$

A converse relation $\breve{R}$ has exactly the same number of ordered pairs as R. So, in particular, when $|X| = n$,

(33) $$|\ni_X| = \sum_{i=0}^{n} \binom{n}{i} i = n2^{n-1},$$

(34) $$\frac{|\ni_X|}{|\mathrm{P}X \times X|} = \frac{n2^{n-1}}{n2^n} = \frac{1}{2},$$

(35) $$|\supset_X| = \sum_{i=0}^{n} \binom{n}{i} 2^i = 3^n,$$

(36) $$\frac{|\supset_X|}{|\mathrm{P}X \times \mathrm{P}X|} = \frac{3^n}{2^{2n}} = \left(\frac{3}{4}\right)^n.$$

1.24 Relative Product Let $R \subset X \times Y$ and $S \subset Y \times Z$ be relations. Their *relative product* (*RL*: 3.8) $S \circ R \subset X \times Z$ is a relation that is the set of all ordered pairs $(x,z) \in X \times Z$ for which there exists an $y \in Y$ with $(x,y) \in R$ and $(y,z) \in S$:

$$S \circ R = \{(x,z) \in X \times Z : \exists y \in Y \quad (x,y) \in R \wedge (y,z) \in S\}. \tag{37}$$

Consider the triple product $X \times Y \times Z$ equipped with its canonical projections $\pi_{12} : X \times Y \times Z \to X \times Y$, $\pi_{13} : X \times Y \times Z \to X \times Z$, and $\pi_{23} : X \times Y \times Z \to Y \times Z$ (cf. Section 1.16). Then $\pi_{12}^{-1}(R) = R \times Z$ and $\pi_{23}^{-1}(S) = X \times S$, whence

$$\begin{aligned} \pi_{12}^{-1}(R) \cap \pi_{23}^{-1}(S) &= (R \times Z) \cap (X \times S) \\ &= \{(x,y,z) \in X \times Y \times Z : (x,y) \in R \wedge (y,z) \in S\} \\ &\subset X \times Y \times Z. \end{aligned} \tag{38}$$

The projection $\pi_{13} : X \times Y \times Z \to X \times Z$ simply eliminates the intermediary 'y-component', thus one has

$$\begin{aligned} \pi_{13}\left(\pi_{12}^{-1}(R) \cap \pi_{23}^{-1}(S)\right) &= \pi_{13}((R \times Z) \cap (X \times S)) \\ &= \{(x,z) \in X \times Z : \exists y \in Y \quad (x,y) \in R \wedge (y,z) \in S\} \\ &= S \circ R. \end{aligned} \tag{39}$$

One may readily verify that relative product is an associative operation: for relations $Q \subset W \times X$, $R \subset X \times Y$, and $S \subset Y \times Z$,

$$(S \circ R) \circ Q = S \circ (R \circ Q) \subset W \times Z \tag{40}$$

(since $S \circ R \circ Q = \{(w,z) \in W \times Z : \exists x \in X \; \exists y \in Y \quad (w,x) \in Q \wedge (x,y) \in R \wedge (y,z) \in S\}$, in which the 'order of appearance' of the intermediaries $x \in X$ and $y \in Y$ is not important)

1.25 Diagonal The *diagonal* of a set X is the relation

$$\Delta_X = \{(x,x) : x \in X\} \subset X \times X. \tag{41}$$

Since the relation is defined, for $x_1, x_2 \in X$, by

$$(x_1, x_2) \in \Delta_X \quad \text{iff} \quad x_1 = x_2, \tag{42}$$

it is also called the *equality* (or *identity*) *relation* on X, and denoted 1_X.

The reader may have noticed that I have already used the symbol 1_X for the identity *map* on X (Definition 1.2). The equivalence between the mapping $1_X : X \to X$ and the relation $1_X \subset X \times X$ will be explained in the general context in the next chapter. Likewise, for $A \subset X$, the inclusion map $i_{A \subset X} : A \to X$ of A in X corresponds to the 'diagonal inclusion' relation

(43) $$\Delta_{A \subset X} = \{(x,x) : x \in A\} \subset A \times X.$$

Note that $\Delta_{A \subset X}$ and $\Delta_A = \{(x,x) : x \in A\} \subset A \times A$ are different relations. Although they have identical graphs and identical domains, $\mathrm{dom}(\Delta_{A \subset X}) = \mathrm{dom}(\Delta_A) = A$, they have different codomains, $\mathrm{cod}(\Delta_{A \subset X}) = X$ but $\mathrm{cod}(\Delta_A) = A$. In the notation of Definition 1.17, the two relations are $(A, X, \Delta_{A \subset X})$ and (A, A, Δ_A).

1.26 Restrictions A relation may be restricted to a subset of its domain or codomain. The *restriction* of the relation $R \subset X \times Y$ to $A \subset X$ is the relation

(44) $$R|_A = R \cap (A \times Y) = R \cap \pi_1^{-1}(A) = \{(x,y) \in R : x \in A\} \subset A \times Y,$$

with domain A and codomain Y. The *restriction* of the relation $R \subset X \times Y$ to $B \subset Y$ is the relation

(45) $$R|^B = R \cap (X \times B) = R \cap \pi_2^{-1}(B) = \{(x,y) \in R : y \in B\} \subset X \times B,$$

with domain X and codomain B. (The canonical projections $\pi_1 : X \times Y \to X$ and $\pi_2 : X \times Y \to Y$ are as in Definition 1.16.)

Note that the corange and range of a restriction are restricted accordingly. The components that are directly restricted are

(46) $$\mathrm{cor}\left(R|_A\right) = \mathrm{cor}(R) \cap A \quad \text{and} \quad \mathrm{ran}\left(R|^B\right) = \mathrm{ran}(R) \cap B.$$

The restrictions are indirect for the duals. For example, suppose $a \in A$ is R-related to $b \in Y \sim B$ but to no other elements of B; i.e., $(a,b) \in R$, and for all other $y \in Y$ $(y \neq b)$, $(a,y) \notin R$. Since $b \notin B$, $b \notin \mathrm{ran}\left(R|^B\right)$. By removing b, the only member of Y to which a is R-related, a is removed from the corange of the restriction, so also $a \notin \operatorname{cor}(R|^B)$; whence a is not $R|^B$-related to b. Thus one sees that, although the restriction is direct on one component, it has an indirect pruning effect on the other component. In general, one has

$$\begin{aligned} \operatorname{ran}(R|_A) &= \pi_2(R|_A) = \pi_2(R \cap \pi_1^{-1}(A)) \\ &= \{y \in Y : \exists x \in A \quad (x,y) \in R\} \subset \operatorname{ran}(R) \end{aligned} \tag{47}$$

and

$$\begin{aligned} \operatorname{cor}(R|^B) &= \pi_1(R|^B) = \pi_1(R \cap \pi_2^{-1}(B)) \\ &= \{x \in X : \exists y \in B \ (x,y) \in R\} \subset \operatorname{cor}(R), \end{aligned} \tag{48}$$

and the inclusions may be proper. One may also characterize $\operatorname{ran}(R|_A)$ as the set of all right R-relatives of elements of A, and $\operatorname{cor}(R|^B)$ as the set of all left R-relatives of elements of B (Definition 1.18).

Domain and codomain restrictions are related through the converse operation:

$$R|_A = \left(\breve{R}|^A\right)^{\smile} \quad \text{and} \quad R|^B = \left(\breve{R}|_B\right)^{\smile}. \tag{49}$$

Alternatively, one may say that the converse operation 'commutes dually' with restriction:

$$(R|_A)^{\smile} = \breve{R}|^A \quad \text{and} \quad (R|^B)^{\smile} = \breve{R}|_B. \tag{50}$$

Restrictions may be defined equivalently as relative products with the diagonal inclusion relation:

$$R|_A = R \circ \Delta_{A \subset X} \subset A \times Y \quad \text{and} \quad R|^B = \breve{\Delta}_{B \subset Y} \circ R \subset X \times B \tag{51}$$

(where the 'converse diagonal inclusion relation' $\breve{\Delta}_{B \subset Y} = \{(y,y) : y \in B\} \subset Y \times B$).

1.27 Extensions For a relation $R \subset X \times Y$, an *extension* of R to a larger domain $A \supset X$ is a relation $S \subset A \times Y$ such that $R = S|_X$, and an *extension* of R to a larger codomain $B \supset Y$ is a relation $S \subset X \times B$ such that $R = S|^Y$. An extension to a larger domain is also called a *continuation* (cf. *ML*: Praefatio).

R and $A \subset X$ intrinsically determine $R|_A$, while R and $B \subset Y$ intrinsically determine $R|^B$; so one says *the* restriction. But R can have more than one extension S to a superset $A \supset X$ (respectively $B \supset Y$). All that is required is for S and R to coincide in $X \times Y$, and S may arbitrarily contain members (in the sense that they are not dictated by R) of $(A \sim X) \times Y$ (respectively $X \times (B \sim Y)$); so one may only say *an* extension rather than *the* extension. In particular, note that an extension of the restriction $R|_A$ or $R|^B$ (back respectively to X or Y) is not necessarily $R \subset X \times Y$ itself (another illustrative example of non-invertiblility; cf. Section 1.22).

Rel

1.28 The Category Rel The category in which the collection of objects is the collection of all sets (in a suitably naive universe of small sets) and where morphisms are relations (as in Definitions 1.17 and 1.18) is denoted **Rel**. Given two sets X and Y, the hom-set **Rel**(X, Y) of *all* relations between X and Y is thus the power set $\mathrm{P}(X \times Y)$.

For relations $R \subset X \times Y$ and $S \subset Y \times Z$, their composite is their *relative product* $S \circ R \subset X \times Z$ (Definition 1.24). The requisite identity morphism in **Rel**(X,X) is the diagonal relation Δ_X (= the equality relation 1_X; Definition 1.25). Note that $\varnothing = 1_\varnothing \in \mathbf{Rel}(\varnothing, \varnothing)$.

1.29 Converse Functor The *converse functor* is the contravariant functor $\mathfrak{C}$: **Rel** $\to$ **Rel** (or equivalently the covariant functor $\overline{\mathfrak{C}}$: **Rel** $\to$ **Rel**^{op}; *ML*: A.10) that sends a set X to itself and a relation $R \in \mathbf{Rel}(X, Y)$ to its converse $\breve{R} \in \mathbf{Rel}(Y,X)$; viz.

$$(52) \qquad \mathfrak{C} : \begin{cases} X \mapsto X & (X \in \mathfrak{O}\mathbf{Rel}) \\ [R \subset X \times Y] \mapsto \left[\breve{R} \subset Y \times X\right] & (R \in \mathfrak{A}\mathbf{Rel}) \end{cases} .$$

The converse and relative product operations interact thus:

$$(53) \qquad (S \circ R)\breve{} = \breve{R} \circ \breve{S};$$

this is the source of the contravariance, whence

(54) $$\mathsf{C}(S \circ R) = \mathsf{C}(R) \circ \mathsf{C}(S).$$

Since $\breve{1}_X = 1_X$, trivially

(55) $$\mathsf{C}(1_X) = 1_{\mathsf{C}X} = 1_X.$$

Also, the involution (30) implies that

(56) $$\mathsf{C} \circ \mathsf{C} = I_{\mathbf{Rel}}$$

(where $I_{\mathbf{Rel}}$ is the identity functor on **Rel**; cf. Section 0.13 and *ML*: A.12(i)) and C is, naturally, itself an involution (and its own inverse functor). $\mathsf{C} : \mathbf{Rel} \to \mathbf{Rel}$ is thus an isomorphism of categories and is a bijection both on objects and on morphisms (Section 0.13).

2 Solus Mappings

— Josquin Desprez (c. 1450–1521)
Tu solus qui facis mirabilia

Unigenitum

A mapping is a special kind of relation. (I have, of course, assumed all along that the reader knows what a mapping is, and is familiar with its associated concepts such as injection, image map, etc. Their redefinition herein is for the purpose of placement in the context of relations, and provides another example of the sometimes unavoidable ouroboros of exposition, as explained in Nota bene.)

2.1 Mapping A *mapping* f is a relation $(X,\ Y,\ G_f)$ (whence $G_f \subset X \times Y$; cf. Definition 1.17) with the following properties:

i. if $(x, y) \in f$ and $(x, z) \in f$, then $y = z$;
ii. $\mathrm{cor}(f) = \mathrm{dom}(f) = X$ (cf. Definition 1.21).

A. H. Louie, *Intangible Life*, Anticipation Science 2,
DOI 10.1007/978-3-319-65409-6_3

The traditional conception of a mapping is that of a processor that assigns to each element of a given set a definite element of another given set. Note that this condition is more than Property 2.1i, which by itself does not preclude the possibility that for some particular $x \in X$ there are no elements $y \in Y$ for which $(x,y) \in f$. Property 2.1ii is an additional requirement on f (*RL*: 1.24). Since the corange of a relation is a subset of the domain, the requirement $\mathrm{cor}(f) = \mathrm{dom}(f)$ says that the relative complement is empty: $\mathrm{dom}(f) \sim \mathrm{cor}(f) = \varnothing$. This makes the entailment $f \mapsto \mathrm{dom}(f)$ intrinsic (cf. Section 1.19).

2.2 Notations For a mapping $f \subset X \times Y$, one says that f is a *mapping of X into Y* (also a *mapping from X to Y*), denoted by

(1) $$f : X \to Y$$

To *each* element $x \in X$, by Definition 2.1, there corresponds a *unique* element $y \in Y$ such that $(x,y) \in f$. The relation between x and y is traditionally denoted by $y = f(x)$ instead of $(x,y) \in f$, and y is called the *value of the mapping f at the element x*. With the $y = f(x)$ notation, the *graph* of f (Definition 1.17) is

(2) $$G_f = \{(x, f(x)) : x \in X\} \subset X \times Y$$

The equivalence of Definitions 1.17 and 1.18 of a relation means that a mapping f is *itself* also a set of ordered pairs, $f \subset X \times Y$. Thus a mapping inherits from a relation the minor equivocation of notation $f = G_f$.

The collection of *all* mappings of X into Y (i.e., mappings with domain X and codomain Y) is a subset of the power set $\mathrm{P}(X \times Y) = \mathbf{Rel}(X, Y)$; this subset is denoted Y^X, and has cardinality $|Y^X| = |Y|^{|X|}$. For finite $|X|$ and $|Y|$, one may see the derivation of this notation thus: for $f : X \to Y$, for each element $x \in X$ there are $|Y|$ possible choices for its value $f(x) \in Y$; there are therefore $|Y|^{|X|}$ possible mappings from X to Y (cf. Section 1.14). $|Y|^{|X|}$ being the cardinality of all mappings of X into Y, it is then natural to denote this set Y^X, and by extension the same symbol is used for all sets X and Y, finite or infinite.

To *trace* (or '*chase*') the path of an element as it is mapped, one uses the 'maps to' arrow and writes

(3) $$f : x \mapsto y$$

(also $f : x \mapsto f(x)$ and $f : [\cdot \mapsto f(\cdot)]$). A mapping thus represented may have its *relational diagram in graph-theoretic* form drawn as

(4)

(*ML*: 5.4–5.11; *RL*: E.6 and 3.1). The hollow-headed arrow denotes the *flow* from input $x \in X$ to output $y \in Y$, and the solid-headed arrow denotes the induction of or constraint upon this flow by the *processor* f. The processor (efficient cause) and output (final cause) relationship may be characterized 'f *entails* y', which may then be denoted using the entailment symbol $\vdash$ (*RL*: 6.1) as

(5) $$f \vdash y$$

Note that the processor f is *that which entails*, and the output y is *that which is entailed*.

For a mapping $f : x \mapsto y$, 'that which is entailed' $\vdash y$ may take on a secondary role, when f *composes* with another mapping. In the *sequential composite* $g \circ f$ (*ML*: 5.13), the output y of f is used as input (material cause) by another mapping $g : y \mapsto z$ (in the material relay $x \mapsto y \mapsto z$), whence $\vdash y$ is called *material entailment* (*RL*: 6.10). In the *hierarchical composite* $f \vdash y \vdash$ (*ML*: 5.14), the output y of f is itself (the efficient cause of) a mapping $y : u \mapsto v$ (i.e., that which is entailed is a functional process), whence $\vdash y$ is called *functional entailment* (*RL*: 6.14). In both compositions, the final cause y of one mapping participates in further entailment involving other mappings.

I do not need the hierarchical composite (functional entailment being a subtitular topic of *RL*) just yet, but let me repeat the definition of sequential composite.

2.3 Composition If $f : X \to Y$ and $g : Y \to Z$ are mappings, then $\operatorname{ran}(f) \subset Y = \operatorname{dom}(g)$, so, for each $x \in X$, the notation $g(f(x))$ makes logical sense as the trace $f : x \mapsto y$ followed by $g : y \mapsto z$, and the value $z = g(y) = g(f(x))$ is well-defined. The mapping $h : X \to Z$ defined for each $x \in X$ by

$$h(x) = g(f(x)) \tag{6}$$

is called the (*sequential*) *composite* of the mappings f and g, and is denoted by $g \circ f$.

As the relation $g \circ f \subset X \times Z$, the composite coincides with the relative product of $f \subset X \times Y$ and $g \subset Y \times Z$: the value $y = f(x) \in Y$ is the 'intermediary' linking $(x, y) \in f$ and $(y, g(y)) \in g$, whence $(x, g(y)) = (x, g(f(x))) \in g \circ f$ (cf. *RL*: 6.10, 6.11).

Set

2.4 The Category Set The category in which the collection of objects is the collection of all sets (in a suitably naive universe of small sets) and the morphisms are mappings is denoted **Set**. Given two sets X and Y, the hom-set $\mathbf{Set}(X, Y)$ is the collection Y^X of all mappings from X to Y. The identity map $1_X : x \mapsto x$ (cf. Definition 1.2) is the identity morphism in $\mathbf{Set}(X, X)$.

By convention, one sets $Y^{\varnothing} = \{\varnothing\}$. Thus there is exactly one mapping from the empty set to any set Y, namely the 'empty mapping' $\varnothing$. In particular, $\varnothing = 1_{\varnothing} \in \mathbf{Set}(\varnothing, \varnothing)$.

Note that the empty set $\varnothing$ is indeed a mapping. This is because $\mathrm{dom}(\varnothing) = \varnothing$, and $\varnothing \subset \varnothing \times Y$ is a relation (albeit a set of ordered pairs with no elements; cf. Examples 1.20). In order to show that $\varnothing$ is *not* a mapping, there must be $x \in \varnothing$ and $y, z \in Y$, such that $(x, y) \in \varnothing$ and $(x, z) \in \varnothing$ but $y \neq z$. Since $\varnothing$ has no elements, this is impossible. All arguments concerning the empty set follow this pattern: to prove that something is *true* about the empty set, one uses the argument that there is *nothing in the empty set to make it false*. In mathematics, a proposition is said to be *vacuously true* when nothing exists to contradict it.

When $X \neq \varnothing$, $\varnothing^X = \varnothing$. This is because if $X \neq \varnothing$, then $f(X) \neq \varnothing$ for any mapping f with $\mathrm{dom}(f) = X$, whence there cannot be any mapping $f : X \to \varnothing$.

Thus, for any set Y, $\mathbf{Set}(\varnothing, Y) = \{\varnothing\}$ whence $|\mathbf{Set}(\varnothing, Y)| = 1$, and for $X \neq \varnothing$, $\mathbf{Set}(X, \varnothing) = \varnothing$ whence $|\mathbf{Set}(X, \varnothing)| = 0$. Indeed, formally, $0 = \varnothing$ and

$1 = \{0\} = \{\varnothing\}$ (Definition 1.4). Since $|\mathbf{Set}(X, Y)| = |Y^X| = |Y|^{|X|}$, in terms of exponents, these cardinality statements translate to, for $n \geq 0$, $n^0 = 1$, and for $m > 0$, $0^m = 0$.

2.5 Graph Functor The *graph functor* $G : \mathbf{Set} \to \mathbf{Rel}$ is

(7) $$G : \begin{cases} X \mapsto X & (X \in \mathbf{O}\,\mathbf{Set}) \\ [f : X \to Y] \mapsto [f \subset X \times Y] & (f \in \mathbf{A}\,\mathbf{Set}) \end{cases}$$

i.e., G sends a set to itself and sends a mapping f to its graph G_f (Notation 2.2).

Since the relation that is the graph of f, $G_f = \{(x, f(x)) : x \in X\} \subset X \times Y$, is but an equivalent representation of the mapping $f : X \to Y$, the category **Set** of sets and mappings may be considered a *subcategory* (Definition 0.6 and *ML*: A.7) of **Rel** of sets and relations: the objects of the two categories are identical, a **Set**-morphism is a **Rel**-morphism, and compositions as **Rel**-morphisms and as **Set**-morphisms are the same for mappings. Thus the graph functor $G : \mathbf{Set} \to \mathbf{Rel}$ is an *inclusion functor* (Definition 1.19). The inclusion functor of **Set** in **Rel** is the identity on objects (so *a fortiori* $\mathbf{O}\,\mathbf{Set} = \mathbf{O}\,\mathbf{Rel}$), and, in particular, sends the identity map $1_X \in \mathbf{Set}(X,X)$ to the identity relation $1_X \in \mathbf{Rel}(X,X)$. A mapping is a special relation, however, so $\mathbf{Set}(X, Y) \subset \mathbf{Rel}(X, Y)$ is (for $X \neq \varnothing$) a proper inclusion, whence **Set** is not a full subcategory of **Rel**.

2.6 The Sparsity of Mappings To see the extent of 'non-fullness' of **Set** as a subcategory of **Rel**, one notes that the requirement for a subset of $X \times Y$ to qualify as a mapping is quite a stringent one: every element $x \in X$ must be related to exactly one element of $y \in Y$. Most relations, i.e., *generic* members of $\mathcal{P}(X \times Y)$, do not have this 'single-valued' property. The cardinality of the hom-set $\mathbf{Rel}(X, Y)$ is

(8) $$|\mathbf{Rel}(X, Y)| = |\mathcal{P}(X \times Y)| = 2^{|X \times Y|} = 2^{|X||Y|},$$

(cf. Section 1.18) and the cardinality of the hom-set $\mathbf{Set}(X, Y)$ is

(9) $$|\mathbf{Set}(X, Y)| = |Y^X| = |Y|^{|X|}$$

(cf. Sections 2.2 and 2.4). Thus one sees that the proportion of mappings to relations is

$$\frac{|\mathbf{Set}(X,Y)|}{|\mathbf{Rel}(X,Y)|} = \frac{|Y|^{|X|}}{2^{|X||Y|}} = \left(\frac{|Y|}{2^{|Y|}}\right)^{|X|} \tag{10}$$

For any set Y, $\mathbf{Rel}(\varnothing, Y) = \mathbf{Set}(\varnothing, Y) = \{\varnothing\}$, so the proportion (10) is 1 when $X = \varnothing$. If $X \neq \varnothing$, $\mathbf{Rel}(X, \varnothing) = \{\varnothing\}$ and $\mathbf{Set}(X, \varnothing) = \varnothing$, so the proportion (10) is 0 when $X \neq \varnothing$ and $Y = \varnothing$. Save these two extreme cases, if both X and Y are nonempty then the proportion (10) falls in the open interval $(0, 1)$. This is, however, an exceedingly small number (i.e., much closer to 0 than to 1). For example, even for trivially small sets X and Y with $|X| = |Y| = 100$, there are only $|\mathbf{Set}(X,Y)| = 100^{100} = 10^{200}$ mappings among $|\mathbf{Rel}(X,Y)| = 2^{10000} \approx 10^{3010}$ relations, so the proportion of mappings to relations is about 10^{-2810}. Indeed, in the limit, one has

$$\frac{|\mathbf{Set}(X,Y)|}{|\mathbf{Rel}(X,Y)|} \to 0 \quad \text{as} \quad |X| \to \infty \quad \text{or} \quad |Y| \to \infty, \tag{11}$$

which one may term alternatively as '$|\mathbf{Set}(X,Y)|$ is dominated asymptotically by $|\mathbf{Rel}(X,Y)|$'. In this sense, $\mathbf{Set}(X,Y)$ is a *sparse* subset of $\mathbf{Rel}(X,Y)$.

The category **Set** has the pride of place in category theory, since it is often given as the prototypical nontrivial example. I submit that **Rel** is in fact the more fundamental category, containing **Set** as a special, sparse, subcategory. I shall have more to say on this later, in Chapter 5, when I present a new mathematical formulation of Natural Law.

Mappings of Sets

2.7 Image Let f be a mapping from X to Y. If $A \subset X$, the *image* of A under f is defined to be the set $f(A)$ of all elements $f(x) \in Y$ with $x \in A$; i.e.,

$$f(A) = \{f(x) : x \in A\} \subset Y. \tag{12}$$

In this notation, the range of f is $\mathrm{ran}(f) = f(X)$. One may also note that, for all $x \in X, f(\{x\}) = \{f(x)\}$.

Recall (Definition 1.26) the restriction of a relation $R \subset X \times Y$ to $A \subset X$ is the relation $R|_A = \{(x,y) \in R : x \in A\} \subset A \times Y$ with domain A, codomain Y, corange $\mathrm{cor}(R|_A) = \mathrm{cor}(R) \cap A$, and range $\mathrm{ran}(R|_A) = \{y \in Y : \exists\, x \in A\ (x,y) \in R\} \subset \mathrm{ran}(R)$. The same operation applied to f yields the restriction of $f : X \to Y$ to $A \subset X$, which is the mapping $f|_A : A \to Y$ with

$$(13)\qquad \begin{aligned} \mathrm{dom}(f|_A) &= A & \mathrm{cod}(f|_A) &= Y \\ \mathrm{cor}(f|_A) &= A & \mathrm{ran}(f|_A) &= f(A). \end{aligned}$$

(Property 2.1ii for $f : X \to Y$ says $\mathrm{cor}(f) = \mathrm{dom}(f) = X$, so $\mathrm{cor}(f|_A) = \mathrm{cor}(f) \cap A = X \cap A = A = \mathrm{dom}(f|_A)$, thus $f|_A : A \to Y$ also satisfies Property 2.1ii. For the range, $\mathrm{ran}(f|_A) = \{y \in Y : \exists\, x \in A \quad (x,y) \in f\} = \{y \in Y : \exists\, x \in A\ \ y = f(x)\} = f(A)$).

The restriction map may be defined equivalently as a sequential composite with the inclusion map (Definition 1.2):

$$(14)\qquad f|_A = f \circ i_{A \subset X} : A \to Y.$$

Note that the graph of the inclusion map $i_{A \subset X}$ is the diagonal inclusion relation $\Delta_{A \subset X}$ (Definition 1.25).

2.8 Surjection The range $f(X) = \mathrm{ran}(f)$ is a subset of the codomain $Y = \mathrm{cod}(f)$, but they need not be equal. When they are, i.e. when $f(X) = Y$, one says that f is a *mapping of X onto Y*, and that $f : X \to Y$ is *surjective* (or is a *surjection*). Note that every mapping maps onto its range.

If $X \neq \varnothing$ and $Y = \varnothing$, then there are no mappings from X to Y and so *a fortiori* no surjections from X to Y. If $X = \varnothing$ then $f(X) = \varnothing$. The empty mapping $\varnothing : \varnothing \to \varnothing$ thus satisfies the requirement $f(X) = Y$ and is therefore surjective; but if $Y \neq \varnothing$, the empty mapping $\varnothing : \varnothing \to Y$ is not surjective.

2.9 Inverse Image If $B \subset Y, f^{-1}(B)$ denotes the set of all $x \in X$ that f maps into B:

(15) $$f^{-1}(B) = \{x \in X : f(x) \in B\} \subset X,$$

and is called the *inverse image* of B under f.

Because of requisite Property 2.1ii, every $x \in X$ is mapped to a value $f(x) \in Y$, thus $f^{-1}(Y)$ must be all of X. Stated otherwise,

(16) $$f^{-1}(Y) = f^{-1}(\mathrm{cod}(f)) = f^{-1}(\mathrm{ran}(f)) = \mathrm{cor}(f) = \mathrm{dom}(f) = X.$$

The restriction (Definition 1.26) of a relation $R \subset X \times Y$ to $B \subset Y$ is the relation $R|^B = \{(x,y) \in R : y \in B\} \subset X \times B$ with domain X, codomain B, range $\mathrm{ran}(R|^B) = \mathrm{ran}(R) \cap B$, and corange $\mathrm{cor}(R|^B) = \{x \in X : \exists\, y \in B \;\; (x,y) \in R\} \subset \mathrm{cor}(R)$. The same operations may be applied to f, but because of Property 2.1ii, one needs to define the restriction of $f : X \to Y$ to $B \subset Y$ as the mapping $f|^B$: $f^{-1}(B) \to B$ with

(17) $$\begin{aligned} \mathrm{dom}(f|^B) &= f^{-1}(B) \qquad & \mathrm{cod}(f|^B) &= B \\ \mathrm{cor}(f|^B) &= f^{-1}(B) \qquad & \mathrm{ran}(f|^B) &= f(X) \cap B. \end{aligned}$$

The corange of $f|^B$, as inherited from the definition of $\mathrm{cor}(R|^B)$, is $\mathrm{cor}(f|^B) = \{x \in X : \exists\, y \in B \;\; (x,y) \in f\} = \{x \in X : \exists\, y \in B \;\; y = f(x)\} = f^{-1}(B)$. Property 2.1ii requires then that one must also define $\mathrm{dom}(f|^B) = f^{-1}(B)$. For the range, $\mathrm{ran}(f|^B) = \mathrm{ran}(f) \cap B = f(X) \cap B$.

2.10 Partition If $y \in Y, f^{-1}(\{y\})$ is abbreviated to $f^{-1}(y)$, whence

(18) $$f^{-1}(y) = \{x \in X : f(x) = y\}.$$

Note that $f^{-1}(y)$ may be the empty set (when $y \in Y \sim \mathrm{ran}(f)$), or may contain more than one element. This means $f^{-1} : y \mapsto f^{-1}(y)$ is not necessarily a mapping from Y to X. As a relation, however, $f^{-1} = \{(y,x) : f(x) = y\} \subset Y \times X$ is defined for all relations $f \subset X \times Y$, and is the latter's converse relation $\breve{f} \subset Y \times X$ (Definition 1.23).

A pairwise disjoint family of nonempty sets, the union of which is the set X, is called a *partition* of X. The sets in the disjoint family are the *blocks* of the partition (*ML*: 1.16). Every $x \in X$ is in exactly one of these blocks. The collection $\{f^{-1}(y) : y \in \mathrm{ran}(f)\}$ forms a partition of X, the block $f^{-1}(y)$ containing the elements of X that are mapped by f to the value y. The equivalence relation corresponding to this partition is R_f, the *equivalence relation on* X *induced by* f (cf. *ML*: 2.19–2.21), defined, for $x_1, x_2 \in X$, by

$$x_1 R_f x_2 \quad \text{iff} \quad f(x_1) = f(x_2). \tag{19}$$

If $\mathrm{ran}(f) = f(X)$ contains exactly one element (say $b \in Y$), then f is called a *constant mapping* (with constant value b); there is only one block ($f^{-1}(b) = X$) in the corresponding partition of X. The equivalence relation induced by a constant mapping is the universal relation $X \times X$ on X (cf. Section 1.20).

2.11 Injection If, for each $y \in Y, f^{-1}(y)$ consists of at most one element of X, then f is said to be an *injective* (also *one-to-one, 1–1*) *mapping* of X into Y. Other commonly used labels are '$f : X \to Y$ is an *injection*', and '$f : X \to Y$ is an *embedding*'.

If $X \neq \varnothing$ and $Y = \varnothing$, then there are no mappings from X to Y and so *a fortiori* no injections from X to Y. If $X = \varnothing$ then for each $y \in Y, f^{-1}(y) = \varnothing$. Thus, for any set Y, the empty mapping $\varnothing : \varnothing \to Y$ satisfies the requirement to be injective.

2.12 Lemma

i. $f : X \to Y$ *is injective iff for every* $y \in \mathrm{ran}(f), f^{-1}(y)$ *is a singleton set in* X.
ii. $f : X \to Y$ *is surjective iff for every nonempty subset* $B \subset Y, f^{-1}(B)$ *is a nonempty subset of* X.

2.13 Inverse Mapping In view of the equivalence in Lemma 2.12i, when $f : X \to Y$ is injective, an *inverse mapping* $f^{-1} : \mathrm{ran}(f) \to X$ exists (with the mild notational equivocation of each singleton set $f^{-1}(y)$ with the element it contains).

Indeed, as a mapping, f^{-1} is necessarily a *one-to-one* mapping of $\operatorname{ran}(f)$ *onto* $X = \operatorname{dom}(f)$.

For $B \subset Y$, the inclusion map $i_{B\subset Y}: B \to Y$ of B in Y is injective and $\operatorname{ran}(i_{B\subset Y}) = B$; the inverse (restricted to $\operatorname{ran}(i_{B\subset Y}) = B$ as its domain) is the mapping $i_{B\subset Y}^{-1} = 1_B : B \to B$. The restriction of $f : X \to Y$ to $B \subset Y$ may then be defined equivalently as a sequential composite with the inverse inclusion map:

(20) $$f|^B = i_{B\subset Y}^{-1} \circ f : f^{-1}(B) \to B.$$

A mapping and its inverse (when it exists) compose to identity mappings thus:

(21) $$f^{-1} \circ f = 1_X \quad \text{and} \quad f \circ f^{-1} = 1_{f(X)}$$

(but not necessarily $f \circ f^{-1} = 1_Y$). One also has the following simple

2.14 Lemma *Let $f : X \to Y$ and $g : Y \to X$ be mappings. If $g \circ f = 1_X$ then f is injective and g is surjective.*

2.15 Bijection If a mapping $f : X \to Y$ is both one-to-one and onto, i.e. both injective and surjective, then f is called *bijective* (or is a *bijection*), and that the mapping f establishes a *one-to-one correspondence* between the sets X and Y.

When a bijection $f : X \to Y$ exists, the sets X and Y are equipotent (Definition 1.3), whence $|X| = |Y|$, $f^{-1} \circ f = 1_X$, and $f \circ f^{-1} = 1_Y$. Bijections are thus **Set**-isomorphisms as well as **Rel**-isomorphisms (Section 0.5). The empty mapping $\varnothing : \varnothing \to \varnothing$ is bijective.

Pigeonholes

2.16 Lemma *If X and Y are finite sets and the mapping $f : X \to Y$ is injective, then $|X| \leq |Y|$.*

PROOF Since f is a mapping, for each $x \in X$ $|f(x)| = 1$. Since f is an injection, all its values $f(x)$s are distinct, whence

(22) $$|X| = \sum_{x \in X} |f(x)| = \left| \bigcup_{x \in X} f(x) \right| = |f(X)|.$$

Since $\mathrm{ran}(f) = f(X) \subset Y$,

(23) $$|f(X)| \leq |Y|$$

(by Theorem 1.6iii). (21) and (22) together then imply $|X| \leq |Y|$. □

Contrapositively, Lemma 2.16 says that if $|X| > |Y|$, then no mapping from X to Y can be injective; i.e., for each $f : X \to Y$ there are at least two elements of X, $x_1, x_2 \in X$ with $x_1 \neq x_2$, at which f has the same value, $f(x_1) = f(x_2)$. Its analogous statement is one of the most useful principles in mathematics:

2.17 The Pigeonhole Principle (Finite Version) *If m items are put into n containers and m > n, then at least one container must contain more than one item.*

The pigeonhole principle has many equivalents that are often useful in proofs. Among them are:

2.18 Corollaries

i. *If m objects are distributed over n places, and if m > n, then some place receives at least two objects.*
ii. *If m objects are distributed over n places, and if m < n, then some place receives no object.*
iii. *If n objects are distributed over n places in such a way that no place receives more than one object, then each place receives exactly one object.*
iv. *If n objects are distributed over n places in such a way that no place receives no object, then each place receives exactly one object.*

Although most straightforwardly applicable to finite sets (such as placing pigeons into pigeonholes), the Pigeonhole Principle may be extended to infinite sets. In terms of cardinal numbers, one has:

2.19 The Pigeonhole Principle *The codomain of an injective mapping must not be smaller than the domain.*

In this form, however, the principle is tautological, since the meaning of the statement $|X| > |Y|$ on cardinal ordering is exactly that there is no injective mapping from X to Y (Theorem 1.6iv).

2.20 Lemma *For any mapping* $f : X \to Y$, $|X| \geq |f(X)| = |\mathrm{ran}(f)|$.

PROOF For each $y \in f(X) = \mathrm{ran}(f)$, $f^{-1}(y) \neq \varnothing$, whence one may choose an $\varepsilon(y) \in f^{-1}(y)$. Since the collection $\{f^{-1}(y) : y \in \mathrm{ran}(f)\}$ forms a partition of X (cf. Definition 2.10 above), the mapping $\varepsilon : f(X) \to X$ is injective. The Pigeonhole Principle thus gives the desired inequality $|X| \geq |f(X)| = |\mathrm{ran}(f)|$. □

2.21 Corollary *If the mapping* $f : X \to Y$ *is surjective, then* $|X| \geq |Y|$.

PROOF If $f : X \to Y$ is surjective, then $|\mathrm{ran}(f)| = |f(X)| = |Y|$. □

So, in sum, for sets X and Y, there are no injections from X to Y if $|X| > |Y|$, and there are no surjections from X to Y if $|X| < |Y|$; these are nicely consistent with the fact that there are no bijections between X and Y if $|X| \neq |Y|$. I invite the reader to consult Chapter 5 of *RL* (especially *RL*: 5.3) for an alternate formulation of these concepts in terms of the adjacency matrix.

If Y is a finite set the choice mapping ε in the proof of Lemma 2.20 is trivially defined; if Y is infinite, since there is no prescription of how infinitely many choices are to be made, one needs to invoke the

2.22 Axiom of Choice (*ML*: 1.37, *RL*: 0.20 and 1.2) Given a nonempty family $\mathfrak{A}$ of nonempty sets, there is a mapping ε with domain $\mathfrak{A}$ such that, for all $A \in \mathfrak{A}$, $\varepsilon(A) \in A$.

It is quite conventional in mathematics that one explicitly acknowledges when a consequence depends on the Axiom of Choice. There are many set-theoretic generalizations of the finite Pigeonhole Principle 2.17 to the infinite, and some of them are in fact equivalent to the Axiom of Choice [Degen 2000].

2.23 Number of Injections There are no injective mappings from X to Y when $|X| > |Y|$. When $|X| \leq |Y|$, the number of injections f from X to Y is the number of different *ordered* arrangements of an $|X|$-element subset

($\{f(x) : x \in X\}$) of a $|Y|$-element set (Y). When $|X| = m$ and $|Y| = n$ are finite (with $m \leq n$), this number is the well-known 'm-permutations of n' (or 'm th falling power of n') in combinatorics, variously denoted $P(n, m)$, $(n)_m$, or $n^{\underline{m}}$ (among a host of others):

$$(n)_m = \frac{n!}{(n-m)!} = n(n-1)\cdots(n-m+1). \tag{24}$$

If, for $m > n$, one defines $(n)_m = 0$, then $(n)_m$ correctly enumerates injections from an m-element set to an n-element set for all nonnegative integers m and n.

When $X \neq \emptyset$ and $Y = \emptyset$, $|X| = m > |Y| = n = 0$, so $(n)_m = 0$. For $X = \emptyset$, $|X| = m = 0$, so with $|Y| = n$ the count is $(n)_0 = 1$, and the single injection from X to Y is the empty mapping $\emptyset : \emptyset \to Y$.

2.24 Number of Surjections The number of surjective mappings from X to Y is 0 when $|X| < |Y|$. When $|X| = m$ and $|Y| = n$ are finite and $m \geq n$, it is

$$\sum_{k=0}^{n} (-1)^k \binom{n}{k} (n-k)^m, \tag{25}$$

where $\binom{n}{k}$ is, of course, the binomial coefficient, the number of different *unordered* k-element subsets of an n-element set, 'k-combinations of n', also denoted $C(n, k)$. (And the unordered differs from the ordered by the factor $k!$, the number of permutations of the k elements: $k!C(n, k) = P(n, k)$.) Because of the symmetry $\binom{n}{k} = \binom{n}{n-k} = \frac{n!}{k!\,(n-k)!}$, the sum (25) may also be written as

$$\sum_{k=0}^{n} (-1)^{n-k} \binom{n}{k} k^m. \tag{26}$$

Note that the $k = n$ term in (25) and the $k = 0$ term in (26) are zero, so the sum (25) may be taken from 0 to $n - 1$ and (26) from 1 to n; but it is conventional to include both endpoints 0 and n in a sum involving binomial coefficients.

The enumeration (25) is derived as follows. Without loss of generality let Y be the indexed set (*RL*: 0.17) $Y = \{y_1, y_2, \ldots, y_n\}$. For $i = 1, 2, \ldots, n$, let A_i be the set of all mappings f from X to Y that do not containing the element y_i in their ranges; i.e.,

$$\begin{aligned} A_i &= \{f \in \mathbf{Set}(X, Y) : \operatorname{ran}(f) \subset Y \sim \{y_i\}\} \\ &= \{f \in \mathbf{Set}(X, Y) : f^{-1}(y_i) = \varnothing\} \end{aligned} \tag{27}$$

A mapping $f : X \to Y$ is surjective iff, for *every* $y \in Y$, $f^{-1}(y) \neq \varnothing$. So surjections from X to Y are precisely those members of $\mathbf{Set}(X, Y)$ that are *not* in any of the A_i s, i.e., members of the set $\bigcap_{i=1}^{n} (\mathbf{Set}(X, Y) \sim A_i)$, which, by De Morgan's Laws (Theorem 1.10), is $\mathbf{Set}(X, Y) \sim \bigcup_{i=1}^{n} A_i$. By the Inclusion–Exclusion Principle (Theorem 1.11),

$$\begin{aligned} \left| \mathbf{Set}(X, Y) \sim \bigcup_{i=1}^{n} A_i \right| &= |\mathbf{Set}(X, Y)| \\ &\quad - |A_1| - |A_2| - \cdots - |A_n| \\ &\quad + |A_1 \cap A_2| + |A_1 \cap A_3| + \cdots + |A_{n-1} \cap A_n| \\ &\quad - |A_1 \cap A_2 \cap A_3| - |A_1 \cap A_2 \cap A_4| \\ &\qquad - \cdots - |A_{n-2} \cap A_{n-1} \cap A_n| \\ &\quad \vdots \\ &\quad + (-1)^n |A_1 \cap A_2 \cap \cdots \cap A_n|. \end{aligned} \tag{28}$$

The size of the intersection sets appearing in (28) depends only on the number of sets in the intersections and not on which sets appear. For each $i = 1, 2, \ldots, n$, $|A_i|$ is the number of mappings from an m-element set to an $(n-1)$-element set, whence $|A_i| = (n-1)^m$; each $|A_i \cap A_j|$ is the number of mappings from an m-element set to an $(n-2)$-element set, whence $|A_i \cap A_j| = (n-2)^m$; etc. Further, elementary combinatorics reveal that in (28) there are $\binom{n}{1}$ terms of the form $|A_i|$; $\binom{n}{2}$ terms of the form $|A_i \cap A_j|$; etc. — $\binom{n}{k}$ terms involving the intersection of k sets. Note that the last $\binom{n}{n} = 1$ term $|A_1 \cap A_2 \cap \cdots \cap A_n|$ is $0 = 0^m$: no mapping can miss all n elements of Y in its range. Together with the first $\binom{n}{0} = 1$ term in the sum, $|\,\mathbf{Set}(X, Y)\,| = n^m$, the number in (28) is thus exactly that in (25).

2.25 Stirling Partition Numbers Let m and n be nonnegative integers. The number of ways to partition (Definition 2.10) an m-element set into n blocks (nonempty subsets) is a *Stirling partition number*, or *Stirling number of the second kind*, commonly denoted $\left\{ {m \atop n} \right\}$.

The Pigeonhole Principle (cf. Corollary 2.18ii) implies that when $m < n$, $\left\{ {m \atop n} \right\} = 0$. By convention, one defines $\left\{ {0 \atop 0} \right\} = 1$; for $m > 0$, one defines $\left\{ {m \atop 0} \right\} = 0$, and it is evident that $\left\{ {m \atop m} \right\} = 1$.

There are, for example, three ways to partition the three-element set $\{1, 2, 3\}$ into two blocks: $\{\{1\}, \{2, 3\}\}$, $\{\{2\}, \{1, 3\}\}$, and $\{\{3\}, \{1, 2\}\}$; thus $\left\{ {3 \atop 2} \right\} = 3$. In a partition, the ordering of the blocks is not important ($\{\{1\}, \{2, 3\}\}$ is the same as $\{\{2, 3\}, \{1\}\}$). If the blocks are ordered, however, then an 'ordered partition' of an m-element set X into n blocks may be identified as the indexed family (*RL*: 0.17 and 0.18) of inverse images $\{f^{-1}(y_i)\}_{i=1,2,\ldots,n}$ of a surjection f from X to $Y = \{y_1, y_2, \ldots, y_n\}$ (cf. Section 2.24). So $\{\{1\}, \{2, 3\}\}$ corresponds to the mapping $f : \{1, 2, 3\} \to \{y_1, y_2\}$ with $f(1) = y_1$ and $f(2) = f(3) = y_2$, and $\{\{2, 3\}, \{1\}\}$ corresponds to the mapping $g : \{1, 2, 3\} \to \{y_1, y_2\}$ with $g(2) = g(3) = y_1$ and $g(1) = y_2$. The surjections f and g are different mappings.

In the previous section, I have already determined the number of surjective mappings from an m-element set onto an n-element set, when $m \geq n$, given in (25) and (26). This number of surjections is closely related to $\left\{ {m \atop n} \right\}$; all one needs to do to account for the ordering is to multiply the latter by $n!$, the number of permutations of the n blocks. Thus

$$(29)\qquad n!\begin{Bmatrix} m \\ n \end{Bmatrix} = \sum_{k=0}^{n}(-1)^k\binom{n}{k}(n-k)^m = \sum_{k=0}^{n}(-1)^{n-k}\binom{n}{k}k^m.$$

The enumeration of surjections from an m-element set onto an n-element set as $n!\begin{Bmatrix} m \\ n \end{Bmatrix}$ works for all combinations of nonnegative integers m and n, even when $m<n$. In particular, the cases involving $m=0$ or $n=0$ are as follows. If $X \neq \varnothing$ and $Y=\varnothing$, then there are no mappings from X to Y and so a fortiori no surjections from X to Y. This is when $m>0$ and $n=0$, and $0!\begin{Bmatrix} m \\ 0 \end{Bmatrix}=0$. If $X=\varnothing$ then $f(X)=\varnothing$. When $Y=\varnothing$ also, the empty mapping $\varnothing : \varnothing \to \varnothing$ satisfies the requirement $f(X)=Y$ and is therefore surjective; this single mapping is enumerated in $0!\begin{Bmatrix} 0 \\ 0 \end{Bmatrix}=1$. If $X=\varnothing$ but $Y\neq\varnothing$, the empty mapping $\varnothing : \varnothing \to Y$ (it being the only mapping from $\varnothing$ to Y) is not surjective; this is when $n>m=0$, and the surjection count is again $n!\begin{Bmatrix} 0 \\ n \end{Bmatrix}=0$. In this final case, for each k from 0 to n, $(n-k)^m=(n-k)^0=1$, so the sum in (29) is $\sum_{k=0}^{n}(-1)^k\binom{n}{k}$, which by the binomial formula $(x+y)^n=\sum_{k=0}^{n}\binom{n}{k}x^{n-k}y^k$ is $(1-1)^n=0$, and matches the value $n!\begin{Bmatrix} 0 \\ n \end{Bmatrix}=0$.

2.26 Number and Proportion of Bijections When X and Y are finite and $|X|=|Y|=n$, a mapping from X to Y is injective iff it is surjective iff it is bijective. The number of injections from X to Y is given by (24) as $(n)_n=n!$. With $\begin{Bmatrix} n \\ n \end{Bmatrix}=1$ and equality (29), the number of surjections from X to Y according to (25) is also equal (as it should) to $n!$.

The proportion of bijections among all mappings from X to Y is thus $n!/n^n$. For example, if $|X|=|Y|=100$, there are $100! \approx 10^{158}$ bijections among $|\mathbf{Set}(X,Y)|=100^{100}=10^{200}$ mappings, so the proportion of bijections to mappings is about 10^{-42}. In the limit, with Stirling's approximation for factorial

($n! \sim \sqrt{2\pi n}\, n^n e^{-n}$ as $n \to \infty$), one sees that the factor n^n conveniently cancels, and the proportion is

$$\frac{n!}{n^n} \sim \sqrt{2\pi n}\, e^{-n} \to 0 \quad \text{as} \quad n \to \infty \tag{30}$$

Thus, between equipotent sets, the collection of bijections is sparse among mappings, just as mappings are themselves sparse among relations (Section 2.6).

2.27 Proportions of Injections and Surjections Let $|X| = m$ and $|Y| = n$ be finite. With $m \leq n$, the proportion of injections among mappings from X to Y is

$$\frac{(n)_m}{n^m} = \left(\frac{n}{n}\right)\left(\frac{n-1}{n}\right)\cdots\left(\frac{n-m+1}{n}\right), \tag{31}$$

where there are m terms in the product. When $m = 0$ or 1, the proportion (31) is $1/1 = 1$; for $m > 1$, the proportion (31) falls in the open interval $(0, 1)$. With fixed $m > 0$, as $n \to \infty$, one sees that each of the m terms in the product approaches 1 as an asymptotic value, and therefore also does the product itself. This observation may be explained using a stochastic form of the Pigeonhole Principle: with m objects to distribute over n places (the assignment of the m values $f(x)$ in Y, for $x \in X$), the larger n is, the less likely it is that two or more objects have to share the same place; in the limit, it is almost certain that each of the m objects will have its own place.

As $m \to \infty$ in (31) (since injectivity requires $m \leq n$, this means $n \to \infty$ also), the limit of the proportion (31) depends on the relative sizes of m and n. If $n \to \infty$ much faster than $m \to \infty$ (in the precise sense that $m = o(n)$, which means $\frac{m}{n} \to 0$), then the proportion $\frac{(n)_m}{n^m} \to 1$. On the other hand, if the growth rate of m and n are comparable (in the precise sense that $n = O(m)$, which means the relative rate $\frac{n}{m}$ is bounded, that $\limsup \frac{n}{m} < \infty$), then the proportion $\frac{(n)_m}{n^m} \to 0$.

With $m \geq n$, the proportion of surjections among mappings from X to Y is

$$n!\left\{\begin{matrix} m \\ n \end{matrix}\right\} \over n^m \tag{32}$$

In the degenerate case $m = n = 0$, the proportion (32) is $1/1 = 1$; when $m > n = 0$, both the numerator and the denominator in (32) are 0. For $n > 0$, the proportion (32) falls in the open interval $(0, 1)$. With fixed $n > 0$, as $m \to \infty$, the asymptotic value of the Stirling partition numbers is given by

$$\left\{\begin{matrix} m \\ n \end{matrix}\right\} \sim \frac{n^m}{n!}, \tag{33}$$

thus the proportion (32) approaches 1 as an asymptotic value. This observation may again be explained using a stochastic form of the Pigeonhole Principle: with m objects to distribute over n places, the larger m is, the less likely it is that any of the fixed number of places is left empty; in the limit, it is almost certain that each of the n fixed places will be occupied.

As $n \to \infty$ in (32) (since surjectivity requires $m \geq n$, this means $m \to \infty$ also), the limit of the proportion (32) again depends on the relative sizes of m and n (in an analogous fashion to injectivity, but with the roles of m and n reversed). If $m \to \infty$ much faster than $n \to \infty$ (i.e., $n = o(m)$), then the proportion $n!\left\{\begin{matrix} m \\ n \end{matrix}\right\} \Big/ n^m \to 1$. On the other hand, if the growth rate of m and n are comparable (i.e., $m = O(n)$), then the proportion $n!\left\{\begin{matrix} m \\ n \end{matrix}\right\} \Big/ n^m \to 0$.

Note that in this section and the previous, I have calculated the proportions of injections, surjections, and bijections among *mappings*. Their proportions among *relations* remain, of course, sparse, since mappings themselves are sparse among relations (cf. Section 2.6), and these special mappings form subsets of all mappings. Thus, as $m \to \infty$ or $n \to \infty$,

$$\frac{(n)_m}{2^{mn}} \to 0, \quad n!\left\{\begin{matrix} m \\ n \end{matrix}\right\} \Big/ 2^{mn} \to 0, \quad \text{and} \quad \frac{n!}{2^{mn}} \to 0. \tag{34}$$

3
Congeries
Set-Valued Mappings

Pares autem vetere proverbio, cum paribus facillime congregantur.

[Like, according to the old proverb, naturally goes with like.]

— Cicero (44 BC)
Cato Maior de Senectute
III. 7.

Part I of *RL* is a pentateuchal exploration of the algebraic theory of set-valued mappings. It also contains the motivations and other natural philosophical reasons on why I consider them congenial and congenital morphisms for relational biology. The enthused reader is invited to consult *RL* for further details on this much-neglected topic in mathematics. I am not repeating that exploration here in *IL*, but I shall lay down the stepping stones necessary for paths that lead to new vistas.

From Points to Sets

3.1 Definition A *set-valued mapping* from set X to set Y is a mapping $F : X \to \mathrm{P}Y$.

A set-valued mapping may be denoted

(1) $$F : X \multimap Y,$$

such that, for each $x \in X$,

(2) $$F(x) = \{y \in Y : y \in F(x)\} \subset Y.$$

A. H. Louie, *Intangible Life*, Anticipation Science 2,
DOI 10.1007/978-3-319-65409-6_4

Note the *point-to-set* nature of a set-valued mapping, as opposed to 'point-to-point' for a standard mapping in Definition 2.1. This difference, however, is a matter of interpretation, since the value $F(x)$ is an element of (or a 'point' in) $\mathrm{P}Y$; it just so happens that the codomain $\mathrm{P}Y$ (of F as a mapping) is a set of sets, whence $F(x)$ also has the dual role of being a set itself. $F(x)$ may contain more than one element of Y, and it is possible that for some $x \in X$, one has $F(x) = \varnothing$.

I have invented in *RL*: 2.1 the special 'forked arrow' $-\!\sqsubset$ to denote set-valued mappings, in distinction from $\rightarrow$ for a standard (single-valued) mapping. Although I often (but not necessarily always) use capital letters to denote set-valued mappings (e.g., F in $F : X -\!\sqsubset Y$) and lowercase to denote standard mappings (e.g., f in $f : X \rightarrow Y$), the two species of arrows are the determinants that distinguish the formal causes. The same symbolic representations suffice for the other arrow diagrams; context determines the nature of the final cause, whether it is an 'element', a 'set', or some other entity. Thus, for $x \in X$ and $B = F(x) \subset Y$, in the set-valued mapping's element-tracing form, one may write

(3) $$F : x \mapsto B.$$

The processor and output relationship may likewise (as in Section 2.2) be characterized '*F entails B*', which may then be denoted using the entailment symbol $\vdash$ as

(4) $$F \vdash B.$$

The input of F is, as for a standard mapping, still a point $x \in X$, but now the output of the mapping F at the element x is a *set* $B = F(x) \subset Y$. The source (material cause) and the value (final cause) of a set-valued mapping are thus different in kind from each other, they belonging to different hierarchical levels ('point' versus 'set'). The property of 'that which is entailed' is inherited by elements from their containing set: if F entails B, F also entails every member of B. This is the logical statement

(5) $$F \vdash B \quad \Rightarrow \quad \forall y \in B \quad F \vdash y.$$

3.2 Single-Valued Mapping A mapping $f : X \rightarrow Y$ defines a very specialized set-valued mapping $j_f : X -\!\sqsubset Y$ such that, for each $x \in X$, the value

(6) $$j_f(x) = \{f(x)\}$$

is a singleton set. Indeed, one can make the formal definition: a set-valued mapping $F : X \multimap Y$ is called *single-valued* if, for each $x \in X$, $F(x)$ is a singleton set. Conversely, a 'single-valued set-valued mapping' $F : X \multimap Y$, with the property that for each $x \in X$ $|F(x)| = 1$, defines a 'standard' mapping $\varepsilon_F : X \to Y$ by $\varepsilon_F : x \mapsto$ the single element in $F(x)$.

3.3 Dominus The *domain* and *codomain* of the set-valued mapping $F : X \multimap Y$ are respectively the sets X and Y, denoted by

$$\text{dom}(F) = X \tag{7}$$

and

$$\text{cod}(F) = Y. \tag{8}$$

The *corange* and the *range* of F are

$$\text{cor}(F) = \{x \in X : \exists\, y \in Y \;\; y \in F(x)\} = \{x \in X : F(x) \neq \varnothing\}, \tag{9}$$

$$\text{ran}(F) = \{y \in Y : \exists\, x \in X \;\; y \in F(x)\} = \bigcup_{x \in X} F(x). \tag{10}$$

Thus $\text{cor}(F) \subset \text{dom}(F)$ and $\text{ran}(F) \subset \text{cod}(F)$, and both inclusions may be proper.

Definition 3.1 prescribes a set-valued mapping as $F \in \textbf{Set}(X, \wp Y)$; *as a mapping* (as per Definition 2.1), therefore, F would have $\text{dom}(F) = \text{cor}(F) = X$ and $\text{cod}(F) = \wp Y$. Note in particular that $\varnothing \in \wp Y$, so $F(x) = \varnothing$ is a legitimate value for $x \in X$, and the mapping Property 2.1ii, that to each element x of the domain X there corresponds a unique element $F(x) \in \wp Y$, is still satisfied with such an assignment. The set-valued mapping $F : X \multimap Y$ *qua set-valued mapping*, however, shall be treated as a morphism from X to Y (not to $\wp Y$) in a category different from **Set** (to be defined presently); this is the reason for defining $\text{dom}(F) = X$ and $\text{cod}(F) = Y$. A standard (i.e., single-valued) mapping $f : X \to Y$ requires $\text{cor}(f) = \text{dom}(f)$, and unless $\varnothing \in Y$, $f(x) = \varnothing$ is not a legitimate value for $x \in X$. For a set-valued mapping $F : X \multimap Y$, $\text{cor}(F)$ simply serves to partition off those $x \in X$ for which $F(x) \neq \varnothing$ from its complement

(11) $$X \sim \operatorname{cor}(F) = \operatorname{dom}(F) \sim \operatorname{cor}(F) = \{x \in X : F(x) = \varnothing\},$$

which is not necessarily the empty set.

From Sets to Sets

3.4 Image Let F be a set-valued mapping from X to Y. If $A \subset X$, the *image* of A under F, is defined as the set

(12) $$F(A) = \bigcup_{x \in A} F(x) \subset Y.$$

The range of F is the image of the domain X of F, as well as the image of the corange of F, under F:

(13) $$\operatorname{ran}(F) = F(X) = \bigcup_{x \in X} F(x) = \bigcup_{\substack{x \in X \\ F(x) \neq \varnothing}} F(x) = F(\operatorname{cor}(X)).$$

3.5 Sequential Composition Let $F : X \mathrel{-\!\!\sqsubset} Y$ and $G : Y \mathrel{-\!\!\sqsubset} Z$ be set-valued mappings. Their *sequential composite* is the set-valued mapping $G \circ F : X \mathrel{-\!\!\sqsubset} Z$ defined by, for $x \in X$,

(14) $$(G \circ F)(x) = \bigcup_{y \in F(x)} G(y) \subset Z.$$

In *RL*: 3.4, I have also defined a second sequential composition $G \circ F : X \mathrel{-\!\!\sqsubset} Z$, called the *square product*. I shall postpone its reintroduction to Chapter 6 here in *IL* when I discuss side-effects.

One readily verifies that, for every $x \in X$,

(15) $$(G \circ F)(x) = G(F(x)).$$

In the iteration on the right-hand side, F takes the point $x \in X$ to the set $F(x) \subset Y$, then G relays the set $F(x) \subset Y$ to the set $G(F(x)) \subset Z$, the *image* of $F(x)$ under G (as in Definition 3.4). On the left-hand side, the sequential composition $F \circ G$ is a set-valued mapping that combines these two steps into one, taking the point $z \in Z$ to the set $(F \circ G)(z)$ (as defined in (14)). So equality

(15) is not a tautology, but a statement that the two sets on either side, while defined differently, are in fact identical. Stated otherwise, the efficient causes on the two sides of (15) take separate and different paths, but beginning with the same material cause x, they reach the same final cause at the end.

When $|F(x)| = 1$, the union in (14) is taken over a single element, whence the sequential composite of single-valued set-valued mappings coincides with the sequential composition of (standard) mappings.

One may also demonstrate that sequential composite is an associative operation by iteratively applying the defining equation (14): for set-valued mappings $E : W\!\!-\!\!\sqsubset X$, $F : X\!\!-\!\!\sqsubset Y$, $G : Y\!\!-\!\!\sqsubset Z$, and $w \in W$

(16)
$$\begin{aligned}((G \circ F) \circ E)(w) &= \bigcup_{x \in E(w)} (G \circ F)(x) = \bigcup_{x \in E(w)} \left(\bigcup_{y \in F(x)} G(y) \right) \\ &= \bigcup_{y \in \bigcup_{x \in E(w)} F(x)} G(y) = \bigcup_{y \in (F \circ E)(w)} G(y) \\ &= (G \circ (F \circ E))(w)\end{aligned},$$

whence

(17)
$$(G \circ F) \circ E = G \circ (F \circ E) : W\!\!-\!\!\sqsubset Z.$$

3.6 Embedding For $A \subset X$, the set-valued mapping $i : A\!\!-\!\!\sqsubset X$ defined by $i(x) = \{x\}$ for all $x \in A$ is called the *inclusion map* (or the *embedding*) of A in X. The inclusion map of X in X is called the *identity map* on X, denoted 1_X (whence $1_X : x \mapsto \{x\}$). These match their definitions as (single-valued) mappings (cf. Section 1.2).

Categories and Functors of Set-Valued Mappings

3.7 The Category Svm The category in which the collection of objects is the collection of all sets (in a suitably naive universe of small sets) and where morphisms are set-valued mappings is denoted **Svm**. Given two sets X and Y, the hom-set $\mathbf{Svm}(X, Y)$ of *all* set-valued mappings from X to Y, in view of Definition 3.1, is identical to the hom-set $\mathbf{Set}(X, \mathrm{P}Y) = (\mathrm{P}Y)^X$. The cardinality of the hom-set $\mathbf{Svm}(X, Y)$ is thus

$$(18)\quad |\mathbf{Svm}(X,Y)| = |\mathbf{Set}(X,\mathrm{P}Y)| = |(\mathrm{P}Y)^X| = \left(2^{|Y|}\right)^{|X|} = 2^{|X|\,|Y|}$$

(cf. Sections 1.14 and 2.2).

For set-valued mappings $F : X \multimap Y$ and $G : Y \multimap Z$, their composite in **Svm** is their *sequential composite* $G \circ F : X \multimap Z$ (Definition 3.5). The requisite identity morphism in $\mathbf{Svm}(X,X)$ is the identity map 1_X (Definition 3.6).

3.8 Embedding Functor The *embedding functor* j : **Set** → **Svm** is

$$(19)\qquad j : \begin{cases} X \mapsto X & (X \in \mathfrak{O}\,\mathbf{Set}) \\ [f : X \to Y] \mapsto [j_f : X \multimap Y] & (f \in \mathfrak{A}\,\mathbf{Set}) \end{cases},$$

where $j_f : x \mapsto \{f(x)\}$ is as in Definition 3.2. This functor is the identity on the class of objects (the object mapping $j = 1_{\mathfrak{O}\mathbf{Set}} : \mathfrak{O}\,\mathbf{Set} \to \mathfrak{O}\,\mathbf{Svm}$), hence, a fortiori, injective and surjective on objects. It is also injective on arrows (the arrow mapping $j : \mathfrak{A}\,\mathbf{Set} \to \mathfrak{A}\,\mathbf{Svm}$ is injective; Definition 0.8) and faithful (for all $X, Y \in \mathfrak{O}\,\mathbf{Set}$, the mapping $j_{X,Y} : \mathbf{Set}(X,Y) \to \mathbf{Svm}(X,Y)$ is injective; Definition 0.14).

3.9 Remark The faithfulness and injectivity on objects and arrows are properties that j shares with an inclusion functor (Definition 0.15). So, although $\mathbf{Set}(X,Y) \not\subset \mathbf{Svm}(X,Y)$, the embedding functor j : **Set** → **Svm** allows the consideration of the category **Set**, through its isomorphic image $j(\mathbf{Set}) \subset \mathbf{Svm}$, as a 'subcategory' of **Svm**.

Most set-valued mappings are not single-valued, however, so **Set** is not a full subcategory of **Svm**. For $f : X \to Y$, the mapping $j_f : x \mapsto \{f(x)\}$ has a very stringent requirement that each $j_f(x) \subset Y$ is a singleton set; an arbitrary mapping $g \in \mathbf{Set}(X,\mathrm{P}Y)$ can assume its value $g(x)$ in any subset of Y. Thus, for $X, Y \in \mathfrak{O}\,\mathbf{Set}$, the mapping

$$(20)\qquad j_{X,Y} : \mathbf{Set}(X,Y) \to \mathbf{Svm}(X,Y) = \mathbf{Set}(X,\mathrm{P}Y)$$

is injective but not surjective.

3.10 Graph The *graph* of a set-valued mapping $F : X \multimap Y$ is the relation $\Gamma_F \subset X \times Y$ defined by

$$\Gamma_F = \{(x,y) \in X \times Y : y \in F(x)\} = \bigcup_{x \in X} (\{x\} \times F(x)) \tag{21}$$

(*RL*: 2.2). When the relation $\Gamma_F \subset X \times Y$ is considered as an ordered triple (cf. Definition 1.17), the graph of the set-valued mapping F is $\Gamma_F = (X, Y, \Gamma)$, in which Γ is the *graph* of the relation Γ_F. The apparent homonymic usage and inherent notational equivocation $\Gamma = \Gamma(\Gamma_F)$ for 'the graph of the graph of F' is resolved upon understanding that $\Gamma_F \subset X \times Y$ is a 'graph' in both senses.

Note that as a set-valued mapping, $F : X \multimap Y$ has its corange defined by (9), that $\mathrm{cor}(F)$ contains all those $x \in X$ at which $F(x) \neq \varnothing$. The graph $\Gamma_F \subset X \times Y$ of $F : X \multimap Y$, on the other hand, is a relation with its corange defined as in Definition 1.21, that

$$\begin{aligned} \mathrm{cor}(\Gamma_F) &= \{x \in X : \exists\, y \in Y \;\; (x,y) \in \Gamma_F\} \\ &= \{x \in X : \exists\, y \in Y \;\; y \in F(x)\}. \end{aligned} \tag{22}$$

This set contains precisely all those $x \in X$ for which $\exists\, y \in F(x)$, whence $F(x) \neq \varnothing$. So (9) and (22) define the same subset $\mathrm{cor}(\Gamma_F) = \mathrm{cor}(F)$ of $\mathrm{dom}(F) = X$.

When $F : X \multimap Y$ is considered as its defining (single-valued) mapping $F : X \to \mathrm{P}Y$, however, it is a priori the relation (that is its graph; cf. Notation 2.2 and Definition 2.5) $G_F \subset X \times \mathrm{P}Y$, and as such, its corange is, by Definition 1.21,

$$\begin{aligned} \mathrm{cor}(G_F) &= \{x \in X \; : \; \exists\, B \in \mathrm{P}Y \; (x,B) \in G_F\} \\ &= \{(x, B) \; : \; x \in X,\ B = F(x) \in \mathrm{P}Y\}. \\ &= \{(x, F(x)) \; : \; x \in X\} \subset X \times \mathrm{P}Y \end{aligned} \tag{23}$$

But, since $\varnothing \in \mathrm{P}Y$, $\varnothing$ is a legitimate value of the mapping $F : X \to \mathrm{P}Y$. So if $x \in X$ is such that $F(x) = \varnothing$, then $(x, \varnothing) \in G_F$ whence $x \in \mathrm{cor}(G_F)$. This means every $x \in X$ is in the corange of the mapping $F : X \to \mathrm{P}Y$, regardless of whether $F(x) = \varnothing$ or $F(x) \neq \varnothing$. Thus (with a notational imprecision that identifies

$\mathrm{cor}(G_F) = \mathrm{cor}(F))\ \mathrm{cor}(F) = \mathrm{dom}(F) = X$, which is the requirement of Definition 2.1.ii of a mapping.

The assignment $F \mapsto \Gamma_F$ establishes, for each pair of sets X and Y, a mapping $\Gamma_{X,Y} : \mathbf{Svm}(X, Y) \to \mathbf{Rel}(X, Y)$. For a mapping $f : X \to Y$, the graph of its corresponding set-valued mapping $j_f : X -\!\!\sqsubset Y$ coincides with the usual definition in Section 2.2:

$$\Gamma_{j_f} = f \;=\; G_f \;=\; \{(x, f(x)) : x \in X\} \;\subset\; X \times Y. \tag{24}$$

The graph of the identity map $1_X : x \mapsto \{x\}$ is the identity relation $1_X \in \mathbf{Rel}(X, X)$.

3.11 Graphs and Compositions The graph of the sequential composite of set-valued mappings is the relative product of their graphs: for set-valued mappings $F : X -\!\!\sqsubset Y$ and $G : Y -\!\!\sqsubset Z$,

$$\Gamma_{G \circ F} = \Gamma_G \circ \Gamma_F. \tag{25}$$

Statement (25) follows from Definitions 1.24, 3.5, and 3.10:

$$\begin{aligned} \Gamma_G \circ \Gamma_F &= \{(x, z) \in X \times Z \;:\; \exists\, y \in Y \;\; (x, y) \in \Gamma_F \;\wedge\; (y, z) \in \Gamma_G\} \\ &= \{(x, z) \in X \times Z \;:\; \exists\, y \in Y \;\; y \in F(x) \;\wedge\; z \in G(y)\} \\ &= \left\{ (x, z) \in X \times Z \;:\; z \in \bigcup_{y \in F(x)} G(y) \right\} \\ &= \{(x, z) \in X \times Z \;:\; z \in (G \circ F)(x)\} \\ &= \Gamma_{G \circ F}\,. \end{aligned} \tag{26}$$

All these properties of graphs culminate in the definition of the

3.12 Graph Functor The *graph functor* $\Gamma : \mathbf{Svm} \to \mathbf{Rel}$ is

$$\Gamma : \begin{cases} X \mapsto X & (X \in \mathcal{O}\mathbf{Svm}) \\ [F : X -\!\!\sqsubset Y] \mapsto [\Gamma_F \subset X \times Y] & (F \in \mathfrak{A}\mathbf{Svm}) \end{cases}. \tag{27}$$

3.13 Restriction to Singleton Set Recall (Section 1.26) that for a relation $R \in \mathbf{Rel}(X, Y)$ and subset $A \subset X$, the range of the restriction $R|_A$ of R to A is the set of all right R-relatives of elements of A:

(28)
$$\begin{aligned}\operatorname{ran}(R|_A) = \pi_2(R|_A) &= \pi_2(R \cap \pi_1^{-1}(A))\\ &= \{y \in Y : \exists x \in A\ (x,y) \in R\} \subset Y.\end{aligned}$$

Let $x \in X$ and consider the range of the restriction $R|_{\{x\}}$ of R to the singleton set $\{x\}$:

(29)
$$\begin{aligned}\operatorname{ran}\left(R|_{\{x\}}\right) = \pi_2\left(R|_{\{x\}}\right) &= \pi_2(R \cap \pi_1^{-1}(x))\\ &= \{y \in Y : (x,y) \in R\} \subset Y,\end{aligned}$$

which is the set of all right R-relatives of $x \in X$. Now define a set-valued mapping $\Phi_R : X \multimap Y$ with

(30)
$$\Phi_R(x) = \operatorname{ran}\left(R|_{\{x\}}\right).$$

This assignment $R \mapsto \Phi_R$ establishes, for each pair of sets X and Y, a mapping $\Phi_{X,Y} : \mathbf{Rel}(X,Y) \to \mathbf{Svm}(X,Y)$.

The mapping $\Phi_{X,X} : \mathbf{Rel}(X,X) \to \mathbf{Svm}(X,X)$ sends the identity relation $1_X \in \mathbf{Rel}(X,X)$ to the identity map $1_X : x \mapsto \{x\}$ in $\mathbf{Svm}(X,X)$:

(31)
$$\Phi_{1_X}(x) = \operatorname{ran}\left(1_X|_{\{x\}}\right) = \{x\} = 1_X(x).$$

For relations $R \subset X \times Y$ and $S \subset Y \times Z$,

(32)
$$\Phi_{S \circ R} = \Phi_S \circ \Phi_R.$$

Note that on the left-hand side of (32), the composition $S \circ R$ is the relative product of the two relations (Definition 1.24), and on the right-hand side, the composition $\Phi_S \circ \Phi_R$ is the sequential composition of two set-valued mappings (Definition 3.5).

The mappings $\Phi_{X,Y} : \mathbf{Rel}(X,Y) \to \mathbf{Svm}(X,Y)$ and $\Gamma_{X,Y} : \mathbf{Svm}(X,Y) \to \mathbf{Rel}(X,Y)$ (from Definition 3.10) are such that, for $R \in \mathbf{Rel}(X,Y)$ and $F \in \mathbf{Svm}(X,Y)$,

(33) $$\Gamma_{\Phi_R} = R \quad \text{and} \quad \Phi_{\Gamma_F} = F,$$

which one sees from a simple examination of the definitions

(34) $$\begin{aligned} \Gamma_F &= \{(x,y) \in X \times Y : y \in F(x)\}, \\ \Phi_R(x) &= \{y \in Y : (x,y) \in R\}. \end{aligned}$$

The inverse identities (33) are universal (they having the same form for all relations R and set-valued mappings F; cf. *ML*: A.20). So the 'variables' R and F may be dropped, with the inverse identities tersely then written as

(35) $$\Gamma_\Phi = 1 \quad \text{and} \quad \Phi_\Gamma = 1.$$

These identities also say that the mappings $\Phi_{X,Y} : \mathbf{Rel}(X,Y) \to \mathbf{Svm}(X,Y)$ and $\Gamma_{X,Y} : \mathbf{Svm}(X,Y) \to \mathbf{Rel}(X,Y)$ are inverses of each other:

(36) $$\Gamma_{X,Y} \circ \Phi_{X,Y} = 1_{\mathbf{Rel}(X,Y)} \quad \text{and} \quad \Phi_{X,Y} \circ \Gamma_{X,Y} = 1_{\mathbf{Svm}(X,Y)},$$

whence the mappings $\Phi_{X,Y}$ and $\Gamma_{X,Y}$ are bijective, and the hom-sets $\mathbf{Rel}(X,Y)$ and $\mathbf{Svm}(X,Y)$ are equipotent.

In summary, one has thus all the ingredients to define the

3.14 Inverse Graph Functor The *inverse graph functor* $\Phi : \mathbf{Rel} \to \mathbf{Svm}$ is

(37) $$\Phi : \begin{cases} X \mapsto X & (X \in \mathcal{O}\,\mathbf{Rel}) \\ [R \subset X \times Y] \mapsto [\Phi_R : X \mathrel{-\!\!\sqsubset} Y] & (R \in \mathcal{A}\,\mathbf{Rel}) \end{cases}.$$

The graph functor $\Gamma : \mathbf{Svm} \to \mathbf{Rel}$ and the inverse graph functor $\Phi : \mathbf{Rel} \to \mathbf{Svm}$ are both the identity map on objects (so $\mathcal{O}\,\mathbf{Svm} = \mathcal{O}\,\mathbf{Rel}$, and, a fortiori, the object maps $\Gamma : \mathcal{O}\,\mathbf{Set} \to \mathcal{O}\,\mathbf{Svm}$ and $\Phi : \mathcal{O}\,\mathbf{Rel} \to \mathcal{O}\,\mathbf{Svm}$ are bijective), and, in view of the mutual inverse relationships (36), the arrow maps $\Gamma : \mathcal{A}\,\mathbf{Set} \to \mathcal{A}\,\mathbf{Svm}$ and $\Phi : \mathcal{A}\,\mathbf{Rel} \to \mathcal{A}\,\mathbf{Svm}$ are also bijective. The two categories **Rel** and **Svm** are thus *isomorphic* in the category **Cat** (of categories and functors; Definition 0.13 and *ML*: A.15). This fact is succinctly represented by the *commutative* diagram

(38)
$$\mathbf{Svm} \underset{\Gamma}{\overset{\Phi}{\rightleftarrows}} \mathbf{Rel}$$

with

(39)
$$\Gamma \circ \Phi = I_{\mathbf{Rel}} \quad \text{and} \quad \Phi \circ \Gamma = I_{\mathbf{Svm}}.$$

($I_{\mathbf{C}} \in \mathbf{Cat}(\mathbf{C}, \mathbf{C})$ is the identity functor of category $\mathbf{C}$.)

Since **Rel** $\cong$ **Svm**, I shall henceforth (except, of course, when the issue at hand has to do explicitly with the isomorphic functorial relations, as for example in the next chapter) not distinguish between the two and use **Rel** for both categories, and consider 'set-valued mapping' to be synonymous with 'relation'. This identification is, indeed, implicitly taken in *RL*; the equivalent Definitions A and B of set-valued mapping (*RL*: 2.1) are appearing here in *IL* as Definition 1.18 of relation and Definition 3.1 of set-valued mapping. Herein I have just explicated the isomorphic identification formally in category-theoretic terms.

Isomorphism of categories is a rather rare occurrence; much more common is when there is an *equivalence* between categories. The latter will be a major subject that I shall address in Part III here in *IL*.

Power Set Functors

I now have all the ingredients to assemble the power set functor on the category **Rel** of sets and relations.

3.15 Covariant Functor The *covariant power set functor* $\mathfrak{P} : \mathbf{Rel} \to \mathbf{Set}$ assigns to a set X its power set $\mathfrak{P}X$, and assigns to a relation $R \subset X \times Y$ the *mapping* $\mathfrak{P}R : \mathfrak{P}X \to \mathfrak{P}Y$ that sends each $A \subset X$ to the *range* of the restriction $R|_A$ of R to A (which is the set of all right R-relatives of elements of A; Section 1.26):

(40)
$$(\mathfrak{P}R)(A) = \operatorname{ran}\left(R|_A\right) = \{y \in Y : \exists x \in A \ (x, y) \in R\} \subset Y;$$

viz.

(41)
$$\mathfrak{P} : \begin{cases} X \mapsto \mathfrak{P}X & (X \in \mathfrak{O}\,\mathbf{Rel}) \\ [R \subset X \times Y] \mapsto \left[\mathfrak{P}R : A \mapsto \operatorname{ran}\left(R|_A\right)\right] & (R \in \mathfrak{A}\,\mathbf{Rel}) \end{cases}.$$

Recall (Definition 3.13) that the set-valued mapping $\Phi_R \in \mathbf{Svm}(X, Y)$ equivalent to the relation $R \in \mathbf{Rel}(X, Y)$ is defined at $x \in X$ to have the value $\Phi_R(x) \subset Y$ that is the set of all right R-relatives of x. One has therefore the relation

(42) $$\Phi_R(x) = \operatorname{ran}\left(R|_{\{x\}}\right) = (\mathrm{P}R)(\{x\}).$$

The functor $\mathrm{P} : \mathbf{Rel} \to \mathbf{Set}$ is injective but not surjective on objects (Definition 0.8): the assignment $X \mapsto \mathrm{P}X$ for $X \in \mathbb{O}\,\mathbf{Rel}$ is not surjective on $\mathbb{O}\,\mathbf{Set}$. This is because of the fact that not every set is a power set; trivially, if the cardinality of a finite set is not a power of 2, then it cannot be the power set of another set.

The functor $\mathrm{P} : \mathbf{Rel} \to \mathbf{Set}$ is injective on arrows hence faithful, but not full and hence not surjective on arrows (Definition 0.14). The non-fullness is due to the fact that the power set mapping $\mathrm{P}R : \mathrm{P}X \to \mathrm{P}Y$ has certain properties that an *arbitrary* mapping $g : \mathrm{P}X \to \mathrm{P}Y$ of power sets does not necessarily satisfy. For example, for $A,\ B \subset X$,

(43) $$\mathrm{P}R(A \cup B) = \mathrm{P}R(A) \cup \mathrm{P}R(B);$$

in particular, for $x_1,\ x_2 \subset X$,

(44) $$\mathrm{P}R(\{x_1, x_2\}) = \mathrm{P}R(\{x_1\}) \cup \mathrm{P}R(\{x_2\}).$$

A general mapping $g : \mathrm{P}X \to \mathrm{P}Y$ may, however, take arbitrarily values in $\mathrm{P}Y$ at each member of $\mathrm{P}X$ and does not have to preserve the union operation.

Indeed, a power set mapping has the following

3.16 Properties (cf. *RL*: 1.19 & 2.10) Let $\mathrm{P}R : \mathrm{P}X \to \mathrm{P}Y$ and $A, B \subset X$. Then

i. $\mathrm{P}R(\varnothing) = \varnothing;$
ii. $A \subset B \;\Rightarrow\; \mathrm{P}R(A) \subset \mathrm{P}R(B);$
iii. $\mathrm{P}R(A \cup B) = \mathrm{P}R(A) \cup \mathrm{P}R(B);$
iv. $\mathrm{P}R(A \cap B) \subset \mathrm{P}R(A) \cap \mathrm{P}R(B);$
v. $\mathrm{P}R(B \sim A) \supset \mathrm{P}R(B) \sim \mathrm{P}R(A);$
vi. $\mathrm{P}R(X) = \operatorname{ran}(R).$

Note that the power set mapping $\mathsf{P}R : \mathsf{P}X \to \mathsf{P}Y$ is *not* a Boolean algebra homomorphism (*ML*: 3.19) from $\mathsf{P}X$ to $\mathsf{P}Y$. One sees, for example, that Property 3.16iv is not the requisite equality. But if the set-valued mapping $\Phi_R \in \mathbf{Svm}(X, Y)$ is *injective* (which means if $x_1 \neq x_2$ then $\mathsf{P}R(\{x_1\}) \cap \mathsf{P}R(\{x_2\}) = \varnothing$; cf. *RL*: 2.7), then $\mathsf{P}R(A \cap B) = \mathsf{P}R(A) \cap \mathsf{P}R(B)$, in which case $\mathsf{P}R : \mathsf{P}X \to \mathsf{P}Y$ is a Boolean algebra homomorphism from $\mathsf{P}X$ onto $\mathrm{ran}(R) \subset Y$.

3.17 Restriction Functor A functor $F : \mathbf{C} \to \mathbf{D}$ may naturally be restricted to a subcategory $\mathbf{B}$ of it domain $\mathbf{C}$, with $F|_{\mathbf{B}} : \mathbf{B} \to \mathbf{D}$ defined in the obvious way: for each $\mathbf{B}$-object X, $F|_{\mathbf{B}}(X) = FX$, and for each $\mathbf{B}$-morphism $\phi \in \mathbf{B}(X, Y)$, $F|_{\mathbf{B}}(\phi) = F\phi \in \mathbf{D}(FX, FY)$; viz.

(45) $$F|_{\mathbf{B}} : \begin{cases} X \mapsto FX & (X \in \mathfrak{O}\mathbf{B}) \\ [\phi : X \to Y] \mapsto [F\phi : FX \to FY] & (\phi \in \mathfrak{A}\mathbf{B}) \end{cases}.$$

The restriction is formally a composition with the inclusion functor $i : \mathbf{B} \to \mathbf{C}$ (Definition 0.15), $F|_{\mathbf{B}} = F \circ i$. This is analogous to the composition for the restriction map in Section 2.7; indeed, on $\mathbf{B}$-objects one has $F|_{\mathbf{B}} = F \circ i_{\mathfrak{O}\mathbf{B} \subset \mathfrak{O}\mathbf{C}}$ and on $\mathbf{B}$-morphisms one has $F|_{\mathbf{B}} = F \circ i_{\mathfrak{A}\mathbf{B} \subset \mathfrak{A}\mathbf{C}}$. One often simply uses the same symbol for the restriction functor, i.e., $F : \mathbf{B} \to \mathbf{D}$, without further comment.

3.18 Covariant Functor on Set When restricted to the subcategory **Set** of **Rel**, the covariant power set functor $\mathsf{P}|_{\mathbf{Set}} : \mathbf{Set} \to \mathbf{Set}$ is the sequential composite of the graph functor (also inclusion functor) $G : \mathbf{Set} \to \mathbf{Rel}$ (Definition 2.5) with $\mathsf{P} : \mathbf{Rel} \to \mathbf{Set}$:

(46) $$\mathsf{P}|_{\mathbf{Set}} = \mathsf{P} \circ G : \mathbf{Set} \to \mathbf{Rel} \to \mathbf{Set}.$$

It assigns to a set X its power set $\mathsf{P}X$, and assigns to a mapping $f : X \to Y$ the mapping $\mathsf{P}f : \mathsf{P}X \to \mathsf{P}Y$ that sends each $A \subset X$ to (in view of the relation $\mathrm{ran}(f|_A) = f(A)$ established in Section 2.7) its *image* $f(A) \subset Y$ under f; viz.

$$\text{(47)} \qquad \mathrm{P}|_{\mathbf{Set}}: \begin{cases} X \mapsto \mathrm{P}X & (X \in \mathfrak{O}\,\mathbf{Set}) \\ [f : X \to Y] \mapsto [\mathrm{P}f : A \mapsto f(A)] & (f \in \mathfrak{A}\,\mathbf{Set}) \end{cases}.$$

This is, of course, the more familiar form of the covariant power set functor from *ML* and *RL*. Henceforth I shall use the same symbol for the restriction functor, i.e., $\mathrm{P} : \mathbf{Set} \to \mathbf{Set}$ (except when the restriction to the subcategory **Set** of **Rel** needs to be emphasized).

The functor $\mathrm{P} : \mathbf{Set} \to \mathbf{Set}$ is likewise injective but not surjective on objects, and injective on arrows hence faithful, but not full and hence not surjective on arrows, for the same reasons as its $\mathrm{P} : \mathbf{Rel} \to \mathbf{Set}$ extension. The non-fullness of $\mathrm{P} : \mathbf{Set} \to \mathbf{Set}$ has an additional illustration: given a mapping $g : \mathrm{P}X \to \mathrm{P}Y$ of power sets, if a mapping $f : X \to Y$ has to satisfy $g = \mathrm{P}f$, then for $x \in X$, one must have $f(x) = g(\{x\})$; but this requires $g(\{x\}) \subset Y$ to be a singleton set, which is a severe restriction on g.

3.19 Image Map For a relation $R \subset X \times Y$, $\mathrm{P}R : \mathrm{P}X \to \mathrm{P}Y$ likewise sends $A \subset X$ to its image $\Phi_R(A) \subset Y$. This is because (40) as defined is

$$\text{(48)} \qquad \mathrm{P}R(A) = \bigcup_{x \in A} \mathrm{P}R(\{x\}) = \bigcup_{x \in A} \Phi_R(x) \subset Y,$$

precisely the definition of the *image* $\Phi_R(A)$ of A under the set-valued mapping $\Phi_R : X \multimap Y$ (Definition 3.4).

3.20 Power Set Mapping as Relations It is important to note that for the power set functor, whether the domain is **Rel** or restricted to **Set**, the codomain is **Set**, so the power set mapping $\mathrm{P}R : \mathrm{P}X \to \mathrm{P}Y$ (or $\mathrm{P}f : \mathrm{P}X \to \mathrm{P}Y$) is a member of $\mathfrak{A}\,\mathbf{Set}$, i.e., a *single-valued* mapping. And as such, it is the relation $\mathrm{P}R \subset \mathrm{P}X \times \mathrm{P}Y$. Explicitly, it is

$$\text{(49)} \qquad \mathrm{P}R = \left\{ \left(A,\ \mathrm{ran}\left(R|_A\right)\right) \ : \ A \subset X \right\},$$

which may also be considered as $G_{\mathrm{P}R}$, 'the graph of $\mathrm{P}R$' (Notation 2.2).

Since the codomain of $\mathrm{P}R : \mathrm{P}X \to \mathrm{P}Y$ is the power set $\mathrm{P}Y$, $\mathrm{P}R$ may also be interpreted as a set-valued mapping as in Definition 3.1, whence $\mathrm{P}R : \mathrm{P}X \multimap Y$, that, at $A \subset X$, its value is

(50) $$(\mathfrak{P}R)(A) = \operatorname{ran}(R|_A) = \{y \in Y : \exists x \in A \ \ (x,y) \in R\} \ \subset \ Y.$$

And as such, and because of the isomorphism of categories **Rel** $\cong$ **Svm** (Section 3.14), it is also the relation $\mathfrak{P}R \subset \mathfrak{P}X \times Y$:

(51) $$\begin{aligned} \mathfrak{P}R &= \{(A,y) : A \subset X, y \in \operatorname{ran}(R|_A)\} \\ &= \{(A,y) : A \subset X, y \in Y, \exists x \in A \ \ (x,y) \in R\} \end{aligned}.$$

The subtle difference between (49) and (51), i.e., between $\mathfrak{P}R \subset \mathfrak{P}X \times \mathfrak{P}Y$ and $\mathfrak{P}R \subset \mathfrak{P}X \times Y$, will become crucial in the consideration of the graph–power-set adjunction later on our journey.

3.21 Contravariant Functor The *contravariant power set functor* $\overline{\mathfrak{P}} : \mathbf{Rel} \to \mathbf{Set}$ assigns to a set X its power set $\mathfrak{P}X$, and assigns to a relation $R \subset X \times Y$ the *mapping* $\overline{\mathfrak{P}}R : \mathfrak{P}Y \to \mathfrak{P}X$ that sends each $B \subset Y$ to the *corange* of the restriction $R|^B$ of R to B (which is the set of all left R-relatives of elements of B; Section 1.26):

(52) $$(\overline{\mathfrak{P}}R)(B) = \operatorname{cor}(R|^B) = \{x \in X : \exists y \in B \ \ (x,y) \in R\} \subset X;$$

viz.

(53) $$\overline{\mathfrak{P}} : \begin{cases} X \mapsto \mathfrak{P}X & (X \in \mathfrak{O}\,\mathbf{Rel}) \\ [R \subset X \times Y] \mapsto [\overline{\mathfrak{P}}R : B \mapsto \operatorname{cor}(R|^B)] & (R \in \mathfrak{A}\,\mathbf{Rel}) \end{cases}.$$

Recall (Definition 1.29) the contravariant converse functor $\mathbf{C} : \mathbf{Rel} \to \mathbf{Rel}$ that sends a set X to itself and a relation $R \in \mathbf{Rel}(X,Y)$ to its converse $\mathbf{C}R = \breve{R} \in \mathbf{Rel}(Y,X)$. Also recall the relations $\operatorname{ran}(\breve{R}) = \operatorname{cor}(R)$ from Section 1.23 and $(R|^B)^{\smile} = \breve{R}\big|_B$ (for $B \subset Y$) from Section 1.26. Then

$$
\begin{aligned}
((\mathrm{P} \circ \mathrm{C})R)(B) &= \left(\mathrm{P}\breve{R}\right)(B) = \mathrm{ran}\left(\breve{R}\big|_B\right) \\
&= \mathrm{ran}\left(\left(R|^B\right)^{\smile}\right) = \mathrm{cor}\left(R|^B\right) = \left(\overline{\mathrm{P}}R\right)(B),
\end{aligned} \tag{54}
$$

whence one has the functorial composite

$$
\overline{\mathrm{P}} = \mathrm{P} \circ \mathrm{C} : \mathbf{Rel} \to \mathbf{Rel} \to \mathbf{Set}. \tag{55}
$$

3.22 Contravariant Functor on Set When restricted to the subcategory **Set** of **Rel**, the contravariant power set functor is $\overline{\mathrm{P}}\big|_{\mathbf{Set}} = \overline{\mathrm{P}} \circ i : \mathbf{Set} \to \mathbf{Rel} \to \mathbf{Set}$ (henceforth simply denoted $\overline{\mathrm{P}} : \mathbf{Set} \to \mathbf{Set}$ unless the restriction needs to be emphasized) that assigns to a set X its power set $\mathrm{P}X$, and assigns to a mapping $f : X \to Y$ the mapping $\overline{\mathrm{P}}f : \mathrm{P}Y \to \mathrm{P}X$ that sends each $B \subset Y$ to (in view of the relation $\mathrm{cor}\left(f|^B\right) = f^{-1}(B)$ established in Section 2.9) its inverse image $f^{-1}(B) \subset X$ under f; viz.

$$
\overline{\mathrm{P}}\big|_{\mathbf{Set}} : \begin{cases} X \mapsto \mathrm{P}X & (X \in \mathfrak{O}\,\mathbf{Set}) \\ [f : X \to Y] \mapsto \left[\overline{\mathrm{P}}f : B \mapsto f^{-1}(B)\right] & (f \in \mathfrak{A}\,\mathbf{Set}) \end{cases}, \tag{56}
$$

which is, similarly, the more familiar form of the contravariant power set functor.

Both contravariant functors $\overline{\mathrm{P}} : \mathbf{Rel} \to \mathbf{Set}$ and $\overline{\mathrm{P}} : \mathbf{Set} \to \mathbf{Set}$ are injective but not surjective on objects, injective but not surjective on arrows, and faithful but not full, for the same reasons as their covariant counterparts.

3.23 Projections For relation $R \subset X \times Y$ and subsets $A \subset X$ and $B \subset Y$, the values $(\mathrm{P}R)(A)$ and $\left(\overline{\mathrm{P}}R\right)(B)$ of the power set mappings satisfy (as established in Section 1.26) the following equalities involving the canonical projections $\pi_1 : X \times Y \to X$ and $\pi_2 : X \times Y \to Y$:

$$
\begin{aligned}
(\mathrm{P}R)(A) &= \mathrm{ran}\left(R|_A\right) = \pi_2\left(R|_A\right) = \pi_2\left(R \cap \pi_1^{-1}(A)\right) \\
\left(\overline{\mathrm{P}}R\right)(B) &= \mathrm{cor}\left(R|^B\right) = \pi_1\left(R|^B\right) = \pi_1\left(R \cap \pi_2^{-1}(B)\right)
\end{aligned} \tag{57}
$$

4
Coniunctio
Functorial Connections

Nonsense and beauty have close connections — closer connections than Art will allow.

— E. M. Forster (1907)
The Longest Journey
Part 1, Chapter XII

Covariance

4.1 Functorial Commutativity It is now time to take stock of the (covariant) functors encountered so far on this journey.

Graph functor (2.5) $G : \mathbf{Set} \to \mathbf{Rel}$

(1)
$$G : \begin{cases} X \mapsto X & (X \in \mathfrak{O}\mathbf{Set}) \\ [f : X \to Y] \mapsto [f \subset X \times Y] & (f \in \mathfrak{A}\mathbf{Set}) \end{cases};$$

Covariant power set functor on **Rel** (3.15) $\mathrm{P} : \mathbf{Rel} \to \mathbf{Set}$

(2)
$$\mathrm{P} : \begin{cases} X \mapsto \mathrm{P}X & (X \in \mathfrak{O}\mathbf{Rel}) \\ [R \subset X \times Y] \mapsto [\mathrm{P}R : A \mapsto \mathrm{ran}(R|_A)] & (R \in \mathfrak{A}\mathbf{Rel}) \end{cases};$$

Covariant power set functor on **Set** (3.18)

$$\mathrm{P}|_{\mathbf{Set}} = \mathrm{P} \circ G : \mathbf{Set} \to \mathbf{Rel} \to \mathbf{Set}$$

(3)
$$\mathrm{P}|_{\mathbf{Set}} : \begin{cases} X \mapsto \mathrm{P}X & (X \in \mathfrak{O}\mathbf{Set}) \\ [f : X \to Y] \mapsto [\mathrm{P}f : A \mapsto f(A)] & (f \in \mathfrak{A}\mathbf{Set}) \end{cases};$$

A. H. Louie, *Intangible Life*, Anticipation Science 2,
DOI 10.1007/978-3-319-65409-6_5

Embedding functor (3.8) $j : \mathbf{Set} \to \mathbf{Svm}$

(4) $$j : \begin{cases} X \mapsto X & (X \in \mathfrak{O}\mathbf{Set}) \\ [f : X \to Y] \mapsto [j_f : X -\!\sqsubset Y] & (f \in \mathfrak{A}\mathbf{Set}) \end{cases};$$

Graph functor (3.12) $\Gamma : \mathbf{Svm} \to \mathbf{Rel}$

(5) $$\Gamma : \begin{cases} X \mapsto X & (X \in \mathfrak{O}\mathbf{Svm}) \\ [F : X -\!\sqsubset Y] \mapsto [\Gamma_F \subset X \times Y] & (F \in \mathfrak{A}\mathbf{Svm}) \end{cases};$$

Inverse graph functor (3.14) $\Phi : \mathbf{Rel} \to \mathbf{Svm}$

(6) $$\Phi : \begin{cases} X \mapsto X & (X \in \mathfrak{O}\,\mathbf{Rel}) \\ [R \subset X \times Y] \mapsto [\Phi : X -\!\sqsubset Y] & (R \in \mathfrak{A}\,\mathbf{Rel}) \end{cases}$$

Their relationships are summarized in the following commutative diagram:

(7)

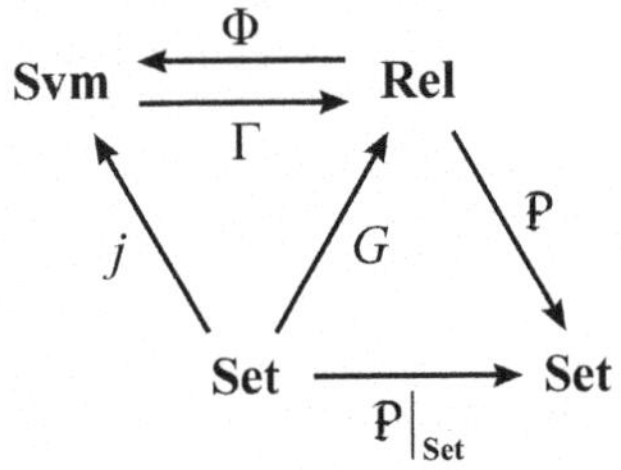

The commuting paths in this diagram yield the following functorial equalities.

4.2 Theorem

i. $\Gamma \circ \Phi = \mathrm{I_{Rel}} : \mathbf{Rel} \to \mathbf{Rel}$;
ii. $\Phi \circ \Gamma = \mathrm{I_{Svm}} : \mathbf{Svm} \to \mathbf{Svm}$;
iii. $\Gamma \circ \mathrm{j} = \mathrm{G} : \mathbf{Set} \to \mathbf{Rel}$;
iv. $\Phi \circ \mathrm{G} = \mathrm{j} : \mathbf{Set} \to \mathbf{Svm}$;
v. $\mathfrak{P} \circ \Gamma \circ \mathrm{j} = \mathfrak{P} \circ \mathrm{G} = \mathfrak{P}|_{\mathrm{Set}} : \mathbf{Set} \to \mathbf{Set}$.

4.3 The Covariant Power Set Functor on Svm The composite functor

(8)

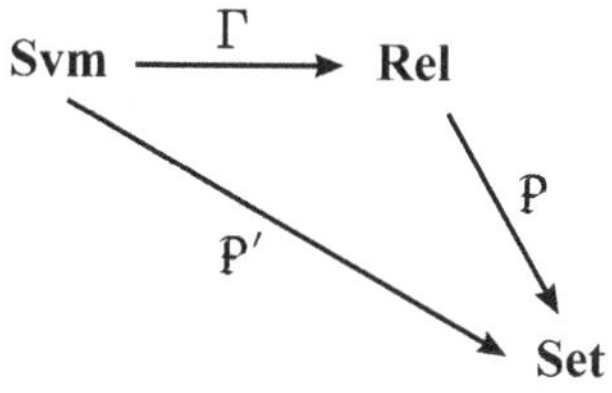

$\mathcal{P}' = \mathcal{P} \circ \Gamma : \mathbf{Svm} \to \mathbf{Rel} \to \mathbf{Set}$ assigns to each set X its power set $\mathcal{P}X$, and sends a set-valued mapping $F : X \mathrel{-\!\!\sqsubset} Y$ to the mapping $\mathcal{P}'F = \mathcal{P}\Gamma_F : \mathcal{P}X \to \mathcal{P}Y$ through the relay

$$(9) \qquad [F : X \mathrel{-\!\!\sqsubset} Y] \mapsto [\Gamma_F \subset X \times Y] \mapsto \left[\mathcal{P}\Gamma_F : A \mapsto \operatorname{ran}\left(\Gamma_F|_A\right)\right].$$

The mapping $\mathcal{P}'F = \mathcal{P}\Gamma_F : A \mapsto \operatorname{ran}\left(\Gamma_F|_A\right)$ is more transparently seen as one that sends each $A \subset X$ to

$$(10) \qquad \begin{aligned} (\mathcal{P}'F)(A) &= \operatorname{ran}\left(\Gamma_F|_A\right) \\ &= \operatorname{ran}(\{(x,y) \in X \times Y : x \in A, y \in F(x)\}) \\ &= \operatorname{ran}\left(\bigcup_{x \in A} (\{x\} \times F(x))\right) \\ &= \bigcup_{x \in A} F(x) = F(A) \in \mathcal{P}Y \end{aligned}$$

Thus the value of the power set mapping $\mathcal{P}'F : \mathcal{P}X \to \mathcal{P}Y$ at $A \in \mathcal{P}X$ is the image of A under F (Definition 3.4).

In summary, the *covariant power set functor* $\mathcal{P}' : \mathbf{Svm} \to \mathbf{Set}$ *on* $\mathbf{Svm}$ is

$$(11) \qquad \mathcal{P}' = \mathcal{P} \circ \Gamma : \begin{cases} X \mapsto \mathcal{P}X & (X \in \mathcal{O}\mathbf{Svm}) \\ [F : X \mathrel{-\!\!\sqsubset} Y] \mapsto [\mathcal{P}'F : A \mapsto F(A)] & (F \in \mathcal{A}\mathbf{Svm}) \end{cases}.$$

Just as in the case of the restriction functor $\mathcal{P} : \mathbf{Set} \to \mathbf{Set}$, I shall henceforth use the same symbol $\mathcal{P} : \mathbf{Svm} \to \mathbf{Set}$ for this power set functor. Since **Svm** and **Rel** are isomorphic, $\mathcal{P} : \mathbf{Svm} \to \mathbf{Set}$ has the same properties as $\mathcal{P} : \mathbf{Rel} \to \mathbf{Set}$ (Section 3.15);

namely, it is injective but not surjective on objects, injective but not surjective on arrows, and faithful but not full.

4.4 Levels The power set functor $\mathfrak{P} : \mathbf{Set} \to \mathbf{Set}$,

(12)
$$\mathfrak{P} : \begin{cases} X \mapsto \mathfrak{P}X & (X \in \mathfrak{O}\mathbf{Set}) \\ [f : X \to Y] \mapsto [\mathfrak{P}f : A \mapsto f(A)] & (f \in \mathfrak{A}\mathbf{Set}) \end{cases},$$

is an essential tool in the analysis of impredicative systems through the reconciliation of two alternate descriptions of an impredicative system S, its 'join of parts' $\vee P$ and its 'whole' W [Louie & Poli 2011].

The entities '$\langle X, f \rangle$' and '$\langle \mathfrak{P}X, \mathfrak{P}f \rangle$' are two alternate descriptions of the same system 'X'. The mapping $f : X \to Y$ maps on the 'element level' (i.e. parts): for each $x \in X$, it assigns an image which is an element $f(x) \in Y$. The mapping $\mathfrak{P}f : \mathfrak{P}X \to \mathfrak{P}Y$ maps on the 'set level' (i.e. whole): for each subset $A \subset X$, it assigns an image which is a subset $\mathfrak{P}f(A) = f(A) \subset Y$. The situation may be represented in the commutative diagram (cf. Section 0.17).

(13)

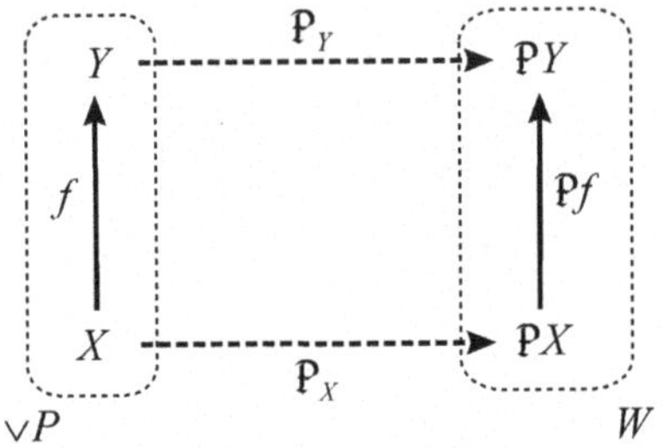

Stated otherwise, for a system S in the category **Set**, the 'join of parts' $\vee P$ is the set as a collection of elements, $X = \{x : x \in X\}$, in a category in which the objects are of the form X, Y, and the morphisms are of the form $f \in H(X, Y)$; the 'whole' W is the set as a lattice of subsets, $X = \sup\{A : A \subset X\} = \bigcup_{A \in \mathfrak{P}X} A$, in a category in which the objects are of the form $\mathfrak{P}X$, $\mathfrak{P}Y$, and the morphisms are of the form $\mathfrak{P}f \in H(\mathfrak{P}X, \mathfrak{P}Y)$. The level-ascending connection is the power set functor $\mathfrak{P} : \mathbf{Set} \to \mathbf{Set}$, in the terse symbolic representation

(14)
$$\mathfrak{P} : \vee P \mapsto W.$$

Contravariance

4.5 Contravariant Functorial Commutativity We have made our acquaintance with these contravariant functors:

Converse functor (1.29) $\mathsf{C} : \mathbf{Rel} \to \mathbf{Rel}$

(15)
$$C : \begin{cases} X \mapsto X & (X \in \mathfrak{O}\mathbf{Rel}) \\ [R \subset X \times Y] \mapsto \left[\breve{R} \subset Y \times X\right] & (R \in \mathfrak{A}\mathbf{Rel}) \end{cases}.$$

Contravariant power set functor on **Rel** (3.21) $\overline{\mathsf{P}} : \mathbf{Rel} \to \mathbf{Set}$

(16)
$$\overline{\mathsf{P}} : \begin{cases} X \mapsto \mathsf{P}X & (X \in \mathfrak{O}\mathbf{Rel}) \\ [R \subset X \times Y] \mapsto \left[\overline{\mathsf{P}}R : B \mapsto \operatorname{cor}\left(R|^{B}\right)\right] & (R \in \mathfrak{A}\mathbf{Rel}) \end{cases};$$

Contravariant power set functor on **Set** (3.22)

$$\overline{\mathsf{P}}\big|_{\mathbf{Set}} = \overline{\mathsf{P}} \circ i : \mathbf{Set} \to \mathbf{Rel} \to \mathbf{Set}$$

(17)
$$\overline{\mathsf{P}}\big|_{\mathbf{Set}} : \begin{cases} X \mapsto \mathsf{P}X & (X \in \mathfrak{O}\mathbf{Set}) \\ [f : X \to Y] \mapsto \left[\overline{\mathsf{P}}f : B \mapsto f^{-1}(B)\right] & (f \in \mathfrak{A}\mathbf{Set}) \end{cases};$$

There is an analogous diagram to (7) when the covariant functors P and $\mathsf{P}|_{\mathbf{Set}}$ are replaced respectively by the contravariant functors $\overline{\mathsf{P}}$ and $\overline{\mathsf{P}}|_{\mathbf{Set}}$:

(18)
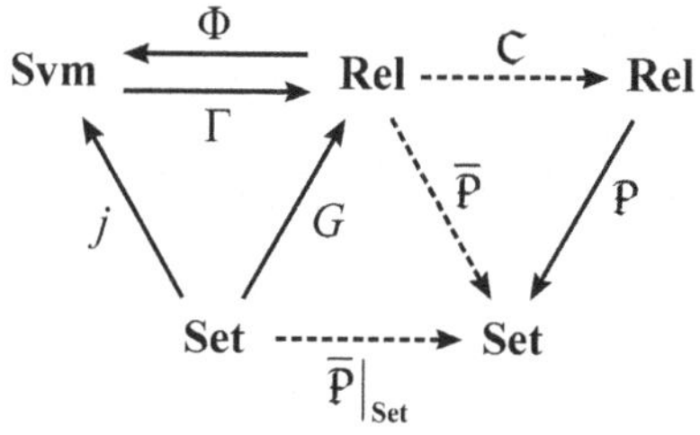

in which I have also added the functorial composite $\overline{\mathsf{P}} = \mathsf{P} \circ \mathsf{C}$ from Section 3.21, and used dashed-line arrows for contravariant functors in distinction from their covariant solid-line arrow counterparts. In this commutative diagram, a path either is a concatenation entirely of solid-line arrows (in which case the functor it represents is covariant) or contains one dashed-line arrow (in which case the functor it represents is contravariant). In the latter, note that all commuting paths (representing the same contravariant functor) also contain one dashed-line arrow.

The commuting paths in diagram (18) yield the following contravariant functorial equalities.

4.6 Theorem

i. $\mathrm{P} \circ \mathrm{C} = \overline{\mathrm{P}} : \mathbf{Rel} \rightarrow \mathbf{Set}.$

ii. $\overline{\mathrm{P}} \circ \Gamma \circ \mathrm{j} = \overline{\mathrm{P}} \circ \mathrm{G} = \overline{\mathrm{P}}\big|_{\mathbf{Set}} : \mathbf{Set} \rightarrow \mathbf{Set}.$

4.7 The Contravariant Power Set Functor on Svm Dual to the covariant power set functor $\mathrm{P} : \mathbf{Svm} \rightarrow \mathbf{Set}$, there is a contravariant power set functor $\overline{\mathrm{P}} : \mathbf{Svm} \rightarrow \mathbf{Set}$. Temporarily denoting it as $\overline{\mathrm{P}'} : \mathbf{Svm} \rightarrow \mathbf{Set}$ for clarity, its composite functorial diagram takes the following form:

(19)

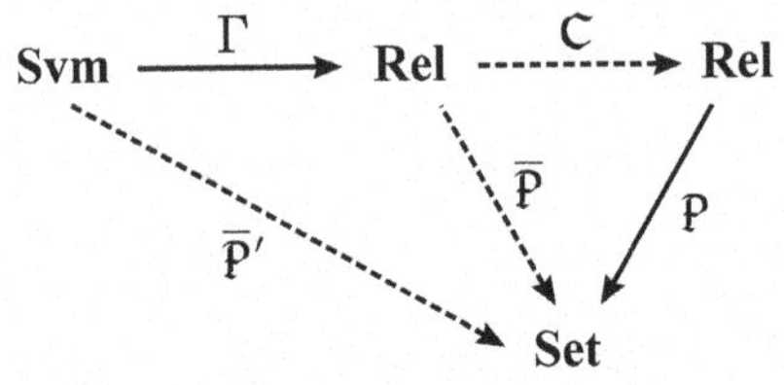

The composite contravariant functor $\overline{\mathrm{P}'} = \overline{\mathrm{P}} \circ \Gamma = \mathrm{P} \circ \mathrm{C} \circ \Gamma : \mathbf{Svm} \rightarrow \mathbf{Rel} \rightarrow \mathbf{Rel} \rightarrow \mathbf{Set}$ assigns to each set X its power set $\mathrm{P}X$. Its action on a set-valued mapping $F : X \mathrel{-\!\!\sqsubset} Y$, however, requires an explanation through the concepts of the inverse mapping of a set-valued mapping and its images (cf. *RL*: 2.13–2.19).

4.8 Inverse Mapping Recall (Definition 1.23) the converse of a relation $R \subset X \times Y$ is the relation $\breve{R} \subset Y \times X$ obtained by interchanging the order of the components, and its graph $\Gamma_{\breve{R}}$ is the transpose of Γ_R:

(20) $$\Gamma_{\breve{R}} = \{(y,x) \in Y \times X : (x,y) \in R\} = [\Gamma_R]^t.$$

For a set-valued mapping $F : X \mathrel{-\!\!\sqsubset} Y$, the converse of its graph $\Gamma_F \subset X \times Y$ is the relation $\breve{\Gamma}_F \subset Y \times X$, which is the graph of the set-valued mapping from

Y to X denoted by $F^{-1} : Y \mathrel{-\!\!\sqsubset} X$, and is called the *inverse* of F. The set-valued mappings F and F^{-1} are related thus:

$$\text{(21)} \qquad \Gamma_{F^{-1}} = \breve{\Gamma}_F \subset Y \times X \quad \text{and} \quad F^{-1} = \Phi_{\breve{\Gamma}_F} : Y \mathrel{-\!\!\sqsubset} X$$

Note that

$$\text{(22)} \qquad \Gamma_{F^{-1}} = \breve{\Gamma}_F = (\mathbf{C} \circ \Gamma)(F).$$

The value of F^{-1} at the point $y \in Y$ is the set

$$\text{(23)} \qquad F^{-1}(y) = \{x \in X : (x, y) \in F\} \subset X,$$

and, for $x \in X$ and $y \in Y$,

$$\text{(24)} \qquad y \in F(x) \quad \text{iff} \quad x \in F^{-1}(y).$$

Inherent from the converse relation (cf. Section 1.23),

$$\text{(25)} \qquad \begin{array}{ll} \operatorname{dom}(F^{-1}) = \operatorname{cod}(F) = Y, & \operatorname{cod}(F^{-1}) = \operatorname{dom}(F) = X, \\ \operatorname{cor}(F^{-1}) = \operatorname{ran}(F) = F(X), & \operatorname{ran}(F^{-1}) = \operatorname{cor}(F) = F^{-1}(Y), \end{array}$$

and

$$\text{(26)} \qquad \left(F^{-1}\right)^{-1} = F.$$

In *RL*: 2.20 and 2.25, I have demonstrated that neither $x \mapsto F^{-1}(F(x))$ nor $y \mapsto F(F^{-1}(y))$ is necessarily the identity mapping on its respective domain; i.e., $F^{-1} \circ F$ is not necessarily the identity map $1_X : X \mathrel{-\!\!\sqsubset} X$ and $F \circ F^{-1}$ is not necessarily the identity map $1_Y : Y \mathrel{-\!\!\sqsubset} Y$. The usage of the term '*inverse* set-valued mapping' appears standard, however; so one must be careful and not equivocate it with the usual algebraic definition in connection with a 'reversal entity for the recovery of the identity'. I have mentioned that some authors call $F^{-1} : Y \mathrel{-\!\!\sqsubset} X$ the 'converse' of $F : X \mathrel{-\!\!\sqsubset} Y$ instead of the 'inverse'. Equation (22) offers the explanation: the graph of F^{-1} is the converse of the graph of F.

4.9 Inverse Image $F^{-1}: Y \multimap X$ is itself a set-valued mapping, so the image $F^{-1}(Y)$ appearing in (25) is already defined as in Definition 3.4: for $B \subset Y$, the image of B under F^{-1} is the set

(27) $$F^{-1}(B) = \bigcup_{y \in B} F^{-1}(y) \subset X.$$

The set $F^{-1}(B)$ is also called the *inverse image* of B by F, and has the equivalent definition as the set

(28) $$F^{-1}(B) = \begin{cases} \{x \in X : F(x) \cap B \neq \varnothing\} & \text{if } B \neq \varnothing \\ \varnothing & \text{if } B = \varnothing \end{cases}$$

(*RL*: 2.16).

4.10 The Path The contravariant functor $\overline{\mathrm{P}'} = \overline{\mathrm{P}} \circ \Gamma = \mathrm{P} \circ \mathrm{C} \circ \Gamma : \mathbf{Svm} \to \mathbf{Rel} \to \mathbf{Rel} \to \mathbf{Set}$ sends a set-valued mapping $F : X \multimap Y$ to the mapping $\overline{\mathrm{P}'}F : \mathrm{P}Y \to \mathrm{P}X$ through the relay

(29) $$\begin{aligned} [F : X \multimap Y] &\mapsto [\Gamma_F \subset X \times Y] \\ &\mapsto \left[\breve{\Gamma}_F = \Gamma_{F^{-1}} \subset Y \times X\right] \\ &\mapsto \left[\mathrm{P}\Gamma_{F^{-1}} : B \mapsto \operatorname{ran}\left(\Gamma_{F^{-1}}|_B\right)\right] \\ &= \left[\mathrm{P}\Gamma_{F^{-1}} : B \mapsto F^{-1}(B)\right]. \end{aligned}$$

Thus the value of the contravariant power set mapping $\overline{\mathrm{P}'}F : \mathrm{P}Y \to \mathrm{P}X$ at $B \in \mathrm{P}Y$ is the inverse image of B by F.

I now revert to the notation $\overline{\mathrm{P}} : \mathbf{Svm} \to \mathbf{Set}$, and summarize the action of the *contravariant power set functor* $\overline{\mathrm{P}} : \mathbf{Svm} \to \mathbf{Set}$ *on* $\mathbf{Svm}$ thus:

(30) $$\overline{\mathrm{P}} : \begin{cases} X \mapsto \mathrm{P}X & (X \in \mathfrak{O}\mathbf{Svm}) \\ [F : X \multimap Y] \mapsto \left[\overline{\mathrm{P}}F : B \mapsto F^{-1}(B)\right] & (F \in \mathfrak{A}\mathbf{Svm}) \end{cases}.$$

This functor is likewise injective but not surjective on objects, injective but not surjective on arrows, and faithful but not full.

Posets Redux

The power set functor $\mathcal{P}$: **Set** $\to$ **Set** is a level-transcending functor; but it so happens that a single power set mapping already contains in itself many functorial properties, in connection with partially ordered sets (posets).

Poset is a major cast member in *ML*. We have already encountered here in *IL*, in Example 0.4iv, that each preordered set and partially ordered set may be considered as a category. For a semblance of completeness (and notational consistence), let me redefine the concepts here.

4.11 Definition If X is a set and $R \subset X \times X$, one says that R is a *relation on X*, and writes xRy instead of $(x, y) \in R$.

4.12 Definition A *preorder* on a set X is relation R on X that is

(r) *reflexive* : for all $x \in X$, xRx; and

(t) *transitive* : for all $x, y, z \in X$, xRy and yRz imply xRz.

One usually uses the notation $\leq$ instead of R when it is a preorder. A *preordered set* (often abbreviated as *proset*) is an ordered pair $\langle X, \leq \rangle$ in which X is a set and $\leq$ is a preorder on X. The universal relation $U = X \times X$ (Section 1.20) is the 'largest' preorder on X, and the diagonal (= equality relation) $\Delta_X = 1_X = \{(x, x) : x \in X\}$ (Definition 1.25) is the 'smallest' preorder on X.

If $x \leq y$ and $y \leq x$, then one writes $x \cong y$ and says that x and y are *isomorphic elements*. Note that $\cong$ is an *equivalence relation* on X, i.e. $\cong$ is (r) reflexive, (t) transitive, and

(s) *symmetric*: for all $x, y \in X$, $x \cong y$ implies $y \cong x$.

A *partial order* $\leq$ is a preorder for which the relation $\cong$ coincides with equality, so that the relation $\leq$ on X is (r) reflexive, (t) transitive, and

(a) *antisymmetric* : for all $x, y \in X$, $x \leq y$ and $y \leq x$ imply $x = y$.

A *partially ordered set* (often abbreviated as *poset*) is an ordered pair $\langle X, \leq \rangle$ in which X is a set and $\leq$ is a partial order on X. The canonical poset is the power set $\langle \mathcal{P}Y, \subset \rangle$ of any set Y.

The converse (Definition 1.23) of a preorder [or partial order] $\leq$ is itself a preorder [or partial order] (Duality Principle; *ML*: 1.26), denoted $\geq$. The *dual* of a p[r]oset $X = \langle X, \leq \rangle$ is the p[r]oset $\tilde{X} = \langle X, \geq \rangle$.

4.13 Definition A mapping f from a p[r]oset $\langle X, \leq_X \rangle$ to a p[r]oset $\langle Z, \leq_Z \rangle$ is called *order-preserving*, *isotone*, or *monotone*, if

(31) $$x \leq_X y \text{ in } X \quad \text{implies} \quad f(x) \leq_Z f(y) \text{ in } Z.$$

An isotone mapping is the requisite structure-preserving morphism in both the category **Pro** of prosets and the category **Pos** of posets.

4.14 Power Set Mapping For any relation $R \in \mathbf{Rel}(X, Y)$ (whence also for the set-valued mapping $\Phi_R \in \mathbf{Svm}(X, Y)$), the power set mapping $\mathrm{P}R = \mathrm{P}\Gamma_{\Phi_R}$: $A \mapsto \mathrm{ran}(R|_A)$ $(A \in \mathrm{P}X)$ is an isotone mapping from the poset $\langle \mathrm{P}X, \subset \rangle$ to the poset $\langle \mathrm{P}Y, \subset \rangle$. This is the statement of Property 3.16ii, that for $A, B \in \mathrm{P}X$,

(32) $$A \subset B \Rightarrow \mathrm{P}R(A) \subset \mathrm{P}R(B).$$

A fortiori, for a mapping $f : X \to Y$, both the covariant power set mapping $\mathrm{P}f$: $A \mapsto f(A)$ and the contravariant power set mapping $\overline{\mathrm{P}}f : B \mapsto f^{-1}(B)$ are isotone (respectively from $\langle \mathrm{P}X, \subset \rangle$ to $\langle \mathrm{P}Y, \subset \rangle$ and vice versa): for $A, B \in \mathrm{P}X$, $A \subset B \Rightarrow f(A) \subset f(B)$; and for $A, B \in \mathrm{P}Y$, $A \subset B \Rightarrow f^{-1}(A) \subset f^{-1}(B)$ (cf. *RL*: 1.19ii and 1.20ii).

The *inclusion* relation $\subset$ on the power set $\mathrm{P}X$ is in fact the canonical partial order, because any poset may be represented as a collection of sets ordered by inclusion:

4.15 Theorem *Let $\langle X, \leq_X \rangle$ be a poset. Define $p : X \to \mathrm{P}X$, for $x \in X$, by*

(33) $$p(x) = \{y \in X : y \leq_X x\}.$$

Then X is isomorphic to the range $p(X) \subset \mathrm{P}X$ of p ordered by set inclusion $\subset$; i.e. $\langle X, \leq_X \rangle \cong \langle p(X), \subset \rangle$.

Note that Theorem 4.15 says that $\langle X, \leq_X \rangle$ and $\langle p(X), \subset \rangle$ are isomorphic posets, i.e., isomorphic objects in the category **Pos**, which is quite different from isomorphic elements within a single p[r]oset in Definition 4.12 above. (The reason for naming the latter thus is explained in the categorical connections below.) In the categories **Pro** and **Pos** (as in any category), to say that two objects X and Y are isomorphic means there are morphisms $f \in H(X, Y)$ and $g \in H(Y, X)$ that are mutual inverses, i.e., $g \circ f = 1_X$ and $f \circ g = 1_Y$. The inverse of $p \in \mathbf{Pos}(\langle X, \leq_X \rangle, \langle p(X), \subset \rangle)$ is $q \in \mathbf{Pos}(\langle p(X), \subset \rangle, \langle X, \leq_X \rangle)$ defined, for $E \subset X$, by

$$q(E) = \sup E. \tag{34}$$

In any concrete category (whence in **Pro** and **Pos** in particular), an isomorphism is a bijection (**Set**-isomorphism). One should note, however, that an isotone bijection is not necessarily a **Pro**-isomorphism. An example is $X = \{a, b\}$ with $\leq_X = \{(a, a), (b, b)\}$ (i.e., $a \leq_X a$ and $b \leq_X b$, but a and b are not $\leq_X$-related to each other); $Y = \{1, 2\}$ with $\leq_Y = \{(1, 1), (1, 2), (2, 2)\}$; and $f : X \to Y$ defined by $f(a) = 1, f(b) = 2$. Then f is an isotone bijective mapping from $\langle X, \leq_X \rangle$ to $\langle Y, \leq_Y \rangle$. The inverse g of f must have the values $g(1) = a$ and $g(2) = b$; but $g : \langle Y, \leq_Y \rangle \to \langle X, \leq_X \rangle$ is not isotone, because $1 \leq_Y 2$ and $g(1) \not\leq_X g(2)$.

4.16 Definition A mapping f from a p[r]oset $\langle X, \leq_X \rangle$ to a p[r]oset $\langle Z, \leq_Z \rangle$ is called *order-reflecting*, or *antitone*, if

$$f(x) \leq_Z f(y) \text{ in } Z \quad \text{implies} \quad x \leq_X y \text{ in } X. \tag{35}$$

4.17 Theorem *A mapping from a p[r]oset to another is an isomorphism if and only if it is bijective, order-preserving (isotone), and order-reflecting (antitone).*

4.18 Imputation Given a mapping $g : X \to Z$ from a set X to a set Z with structure (cf. Section 0.16), it is often possible to impute the mathematical structure of Z back to X by the inverse image mapping $\overline{\mathfrak{P}}g : B \mapsto g^{-1}(B)$ $(B \in \mathfrak{P}Z)$. This is in fact an important consequence of the modelling relation (cf. *ML*: 4.4, 4.5, and 7.37).

In particular, if $\langle Z, \leq_Z \rangle$ is a proset, then one may define a preorder $\leq_X$ on X thus: for $x, y \in X$,

$$\text{if } g(x) \leq_Z g(y) \text{ in } Z \quad \text{then} \quad x \leq_X y \text{ in } X. \tag{36}$$

4.19 Categorical Connections Recall (Example 0.4iv) that a preordered set $\langle X, \leq \rangle$ is a thin category $\mathbf{C}$, in which every hom-set $\mathbf{C}(x, y)$ contains at most one morphism; $\mathfrak{O}\mathbf{C} = X$ and $\mathfrak{A}\mathbf{C} = \leq$. An isotone mapping, which is a preorder-preserving morphism, is then a functor between thin categories.

Suppose $x \leq y$ and $y \leq x$ in the proset $\langle X, \leq \rangle$. That $x \leq y$ means that there is a **C**-morphism $f \in \mathbf{C}(x,y) \neq \varnothing$, and $y \leq x$ means that there is a **C**-morphism $g \in \mathbf{C}(y,x) \neq \varnothing$. These compose to $g \circ f \in \mathbf{C}(x,x)$ and $f \circ g \in \mathbf{C}(y,y)$. But the condition that **C** is thin, that it is the categorical representation of a preorder, means that every hom-set has at most one element. One already has the requisite $1_x \in \mathbf{C}(x,x)$ and $1_y \in \mathbf{C}(y,y)$, so one must have $g \circ f = 1_x$ and $f \circ g = 1_y$, which says precisely that f and g are **C**-isomorphisms (Definition 0.5), whence $x \cong y$, that x and y are *isomorphic* (as **C**-objects). This matches with the usage of the term 'isomorphic elements' in Definition 4.12.

In a partial order, the additional antisymmetry property, that $x \leq y$ and $y \leq x$ imply $x = y$, then corresponds categorically to the situation that whenever any two objects are isomorphic in a skeletal category **C**, they must in fact be equal.

4.20 Isotone Mapping as Functor An isotone mapping f from a proset $\langle X, \leq_X \rangle$ to a proset $\langle Z, \leq_Z \rangle$, when the prosets are considered categories, is then a functor:

(37)
$$f : \begin{cases} x \mapsto f(x) & (x \in X) \\ [x \leq_X y] \mapsto [f(x) \leq_Z f(y)] & (x, y \in X) \end{cases}.$$

For a relation $R \in \mathbf{Rel}(X, Y)$, the power set mapping $\mathrm{P}R : \langle \mathrm{P}X, \subset \rangle \to \langle \mathrm{P}Y, \subset \rangle$ is isotone; it is therefore the functor

(38)
$$\mathrm{P}R : \begin{cases} A \mapsto \mathrm{ran}\left(R|_A\right) & (A \in \mathrm{P}X) \\ [A \subset B] \mapsto \left[\mathrm{ran}\left(R|_A\right) \subset \mathrm{ran}\left(R|_B\right)\right] & (A, B \in \mathrm{P}X) \end{cases}.$$

A fortiori, for a mapping $f : X \to Y$, the power set mappings $\mathrm{P}f : \mathrm{P}X \to \mathrm{P}Y$ and $\overline{\mathrm{P}}f : \mathrm{P}Y \to \mathrm{P}X$ are functors:

(39)
$$\mathrm{P}f : \begin{cases} A \mapsto f(A) & (A \in \mathrm{P}X) \\ [A \subset B] \mapsto [f(A) \subset f(B)] & (A, B \in \mathrm{P}X) \end{cases};$$

(40)
$$\overline{\mathrm{P}}f : \begin{cases} B \mapsto f^{-1}(B) & (B \in \mathrm{P}Y) \\ [A \subset B] \mapsto [f^{-1}(A) \subset f^{-1}(B)] & (A, B \in \mathrm{P}Y) \end{cases}.$$

4.21 Pro(X, Y) as Proset Let f and g be isotone mappings from a proset $\langle X, \leq_X \rangle$ to a proset $\langle Z, \leq_Z \rangle$. Define

$$f \preceq g \quad \text{iff} \quad \text{for all } x \in X \quad f(x) \leq_Z g(x) \tag{41}$$

This establishes a preorder $\preceq$ on the hom-set **Pro**(X, Y) of all isotone mappings from $\langle X, \leq_X \rangle$ to $\langle Z, \leq_Z \rangle$.

4.22 Pro(X, Y) as Functor Category Consider two isotone mappings (which are also functors) $f, g : \langle X, \leq_X \rangle \to \langle Z, \leq_Z \rangle$ between prosets. Each hom-set of the thin category $\langle Z, \leq_Z \rangle$ contains at most one morphism, there is therefore at most one natural transformation τ from f to g (cf. Definition 0.18), since there is at most one way to define the component of τ at $x \in X$, the 'Z-morphism' τ_x from $f(x)$ to $g(x)$. Now the Z-hom-set from $f(x)$ to $g(x)$ is nonempty when $f(x) \leq_Z g(x)$. Thus τ requires, for all $x \in X, f(x) \leq_Z g(x)$; i.e., $f \preceq g$. Stated otherwise, there is a natural transformation from the functor f to the functor g if and only if $f \preceq g$. The commutativity condition ($t1$) for the natural transformation τ is, for $x, y \in X$,

$$\left.\begin{array}{l} f(x) \leq_Z g(x) \leq_Z g(x) \\ f(x) \leq_Z f(y) \leq_Z g(y) \end{array}\right\} \Rightarrow f(x) \leq_Z g(y). \tag{42}$$

4.23 Ordering of Power Set Mappings For two set-valued mappings $F, G : X \multimap Y$, their covariant and contravariant power set mappings, $\mathrm{P}F, \mathrm{P}G : \langle \mathrm{P}X, \subset \rangle \to \langle \mathrm{P}Y, \subset \rangle$ and $\overline{\mathrm{P}}F, \overline{\mathrm{P}}G : \langle \mathrm{P}Y, \subset \rangle \to \langle \mathrm{P}X, \subset \rangle$ respectively, are isotone mappings between posets.

The ordering $\mathrm{P}F \preceq \mathrm{P}G$ is the equivalence that for all $A \subset X$, $F(A) \subset G(A)$. The condition applies, in particular, to $A = \{x\}$ for all $x \in X$. That is, $\mathrm{P}F \preceq \mathrm{P}G$ iff for all $x \in X$, $F(x) \subset G(x)$, which is simply the inclusion $F \subset G$ as relations $F, G \subset X \times Y$.

Contravariantly, the ordering $\overline{\mathrm{P}}F \preceq \overline{\mathrm{P}}G$ is the equivalence that for all $B \subset Y$, $F^{-1}(B) \subset G^{-1}(B)$. The condition applies, in particular, to $B = \{y\}$ for all $y \in Y$. That is, $\overline{\mathrm{P}}F \preceq \overline{\mathrm{P}}G$ iff for all $y \in Y$, $F^{-1}(y) \subset G^{-1}(y)$. When this inclusion holds, if $x \in F^{-1}(y)$ then $x \in G^{-1}(y)$, which says if $y \in F(x)$ then $y \in G(x)$. Thus, again,

$\overline{\mathrm{P}}F \preceq \overline{\mathrm{P}}G$ iff for all $x \in X$, $F(x) \subset G(x)$, which is likewise simply the inclusion $F \subset G$ as relations $F, G \subset X \times Y$.

With two (single-valued) mappings $f, g : X \to Y$, when either of the power set mapping orderings holds, $\mathrm{P}f \preceq \mathrm{P}g$ or $\overline{\mathrm{P}}f \preceq \overline{\mathrm{P}}g$., that $f \subset g$, the consequence is equality $f = g$.

Part II
Sicut
Natural Law and the Modelling Relation

—Philip Van Wilder (c. 1500–1554)
Pater noster

Transcent

Why mathematical biology?

The final cause is biology, the study (with its goal the understanding) of life. The efficient cause is mathematics, the tools of the approach. The material cause is all the objects and relations in mathematics and biology, the source ingredients with which to make connections. The formal cause, i.e. the general premise, of mathematical biology is to represent a biological entity by an appropriate mathematical object. The representation should faithfully reflect the biological properties of the living system in the mathematical properties of the associated formal system, so that different mathematical structures may be compared by appropriate structure-preserving mappings. In short, the formal cause of mathematical biology is category theory, a general theory of modelling.

In Part I, I have explicated in detail the level-ascending power set functor P: **Rel** $\rightarrow$ **Set** and its extended family of functorial connections. The descriptions $\langle X, R\rangle$ and $\langle \mathrm{P}X, \mathrm{P}R\rangle$, and more generically the join of parts $\vee P$ and the whole W, interact and integrate as alternate models of the same system but on different hierarchical levels. In Part II, the theatre is expanded to other transcendent functorial connections, in the modelling relation in general, and in relational biology in particular.

5
Modus
Rational Nature

> Ich habe keinen besseren Ausdruck als den Ausdruck "religiös" für dieses Vertrauen in die vernünftige und der der menschlichen Vernunft wenigstens einigermassen zugängliche Beschaffenheit der Realität. Wo dieses Gefühl fehlt, da artet Wissenschaft in geistlose Empirie aus.
>
> [I have no better expression than 'religious' for this confidence in the rational nature of reality and in its being accessible, to some degree, to human reason. When this feeling is missing, science degenerates into mindless empiricism.].
>
> — Albert Einstein (1951)
> Letter to Maurice Solovine, 1 January 1951

Category theory is a general theory of modelling, and a functor between categories plays the role of a relation between the system being modelled and the mathematical object that is the model. The objects in a given category are metaphors for one another, in the sense that once one knows how to encode any one of them into a model, one knows how to encode all the others. Whenever a modelling relation is established between systems, one has built at least a piece of a functor.

Indeed, the essence of a functor $F : \mathbf{C} \to \mathbf{D}$ is the simultaneous establishment of the sense in which all **C**-objects and **C**-morphisms in the domain category are related to one another and of explicit models of them as **D**-objects and **D**-morphisms in the codomain category. It must be emphasized that a functorial connection between two categories is more than a local correspondence between $A \in \mathfrak{O}\mathbf{C}$ and $FA \in \mathfrak{O}\mathbf{D}$ and between $f \in \mathfrak{A}\mathbf{C}$ and $Ff \in \mathfrak{A}\mathbf{D}$ (and, a fortiori, much more than a lower-level 'pointwise' correspondence in concrete categories). A functorial connection, more importantly, captures certain global features of $\mathfrak{O}\mathbf{C}$ and $\mathfrak{A}\mathbf{C}$, and manifests them in corresponding global properties in $\mathfrak{O}\mathbf{D}$ and $\mathfrak{A}\mathbf{D}$. Thus a functor is in principle not a reductionistic kind of relation.

A. H. Louie, *Intangible Life*, Anticipation Science 2,
DOI 10.1007/978-3-319-65409-6_6

Systems

'*System*' is a primitive. It takes on the intuitive meaning:

5.1 System A *system* is a collection of material or immaterial things that comprises one's subject of study.

5.2 Material System A *material system* is a physical object in the world.

5.3 Natural System (*ML*: 4.4) A *natural system* is
i. a part, whence a subset, of the external world; and
ii. a collection of qualities, to which definite relations can be imputed.

There is a subtle difference between a material system (also called a physical system or a physicochemical system) and a natural system. A material system is ontological, it being simply *any* physical object in the world. A natural system, on the other hand, is epistemological, since the partitioning of the external world and the formation of percepts and their relations are all mental constructs (and are therefore entailed by the bounds of mental constructs). In short, *a natural system is a subjectively-defined representation of a material system* (*ML*: 7.18). In a natural system that comprises one's subject of study, the material is the tangible physical components, and the immaterial is the intangible processes that are causal interactions among the components.

When the 'subject of study' is in the external world of sensory phenomena with their causal entailment, the system is called a natural system. When the object of study is the internal world of ideas with their inferential entailment expressed in some language, the system is called a formal system.

5.4 Definition A (*ML*: 4.6) A *formal system* is an object in the universe of mathematics.

The category of formal systems is the subject of Chapter 7 of *ML*.

5.5 Definition B (*ML*: 7.1; *RL*: 7.8) A *formal system* is a pair $\langle S, F \rangle$, where S is a set, and F is a collection of mappings with domain S, such that $0 \in F$, where 0 is (the equivalence class of) the constant mapping on S.

Recall $0 \in F$ is an algebraic requirement; the role of the constant mapping 0 (cf. Section 2.10) is to identify the set S itself, and to define the property of 'belonging to S'. Indeed, for each $f \in F, \mathrm{dom}(f) = S$; so the collection F of observables implicitly defines the set S, whence a formal system may alternatively be defined as a collection F of mappings with a common domain.

'An object in the universe of mathematics' may, of course, be interpreted as an appropriately defined category **C**. In a formal system that comprises one's subject of study, the collection of 'material things' is the objects $\mathfrak{O}\mathbf{C}$, and the collection of 'immaterial things' is the processes that are inferential interactions among the objects, the morphisms $\mathfrak{A}\mathbf{C}$. The terse Definition 5.4 may then, without loss of generality, take on a more formal mathematical definition, that of a category.

5.6 Definition C A *formal system* is a category $\mathbf{C} = \langle \mathfrak{O}\mathbf{C}, \mathfrak{A}\mathbf{C} \rangle$, comprising a collection $\mathfrak{O}\mathbf{C}$ of *objects* and a collection $\mathfrak{A}\mathbf{C}$ of *arrows* (*morphisms*), satisfying the category-theoretic axioms as in Definition 0.2.

5.7 Observable (cf. *ML*: 2.23) Let $A \in \mathfrak{O}\mathbf{C}$. An *observable* of A is a **C**-morphism with domain A. The collection of all observables of A is the union of all the **C**-hom-sets with domain A, denoted $\mathbf{C}(A, \bullet)$:

(1)
$$\mathbf{C}(A, \bullet) = \bigcup_{B \in \mathfrak{O}\mathbf{C}} \mathbf{C}(A, B) \subset \mathfrak{A}\mathbf{C}.$$

Note that this is a disjoint union, since $\mathbf{C}(A, B_1) \cap \mathbf{C}(A, B_2) = \varnothing$ if $B_1 \neq B_2$.

The collection of **C**-hom-sets $\{\mathbf{C}(A, B) : A, B \in \mathfrak{O}\mathbf{C}\}$, indeed, forms a partition of $\mathfrak{A}\mathbf{C}$ (Definitions 0.1 and 0.2; *RL*: 6.9). This partition is a subpartition of a coarser partition (*ML*: 2.13) formed by joining **C**-hom-sets with a common domain, i.e., the partition of $\mathfrak{A}\mathbf{C}$ into observables indexed by $A \in \mathfrak{O}\mathbf{C}$:

(2)
$$\{\mathbf{C}(A, \bullet) : A \in \mathfrak{O}\mathbf{C}\}.$$

In view of the more general Definition 5.6, the legacy formal system $\langle S, F \rangle$ of Definition 5.5 is more appropriately renamed an *atomic formal system in the category* **Set**. Note that the pair $\langle S, F \rangle$, where $F \subset \mathbf{Set}(S, \bullet)$ (with the notation introduced in (1)), does not by itself contain enough requisite ingredients to be a category. A subcategory **C** of **Set** that contains $\langle S, F \rangle$ would have to contain as a minimum, in addition to $S \in \mathfrak{O}\mathbf{C}$ and $F \subset \mathbf{C}(S, \bullet) \subset \mathfrak{A}\mathbf{C}$, in $\mathfrak{O}\mathbf{C}$ the codomains of all the morphisms in $\mathbf{C}(S, \bullet)$ and in $\mathfrak{A}\mathbf{C}$ the identity morphisms of all the objects in $\mathfrak{O}\mathbf{C}$. Henceforth the term *atomic formal system in the category* **Set** shall be a modified version of Definition 5.5, that $\langle S, F \rangle$ is the *smallest subcategory of* **Set** that contains the set S and the mappings $F \subset \mathbf{Set}(S, \bullet)$.

5.8 The Bull Transcended All is one law, not two. In a system, whether formal or natural, the material and immaterial things may be unified. Function dictates structure; the intangible is the seed of the tangible. Just as mappings and their compositions in formal systems, causal processes in a natural system similarly connect their interacting components. A category **C** may simply be considered a collection of interconnected arrows, with many natural ways to formulate the inferential entailment $\mathfrak{A}\mathbf{C} \vdash \mathfrak{O}\mathbf{C}$. On this level of generality, then, a system in relational, hence non-material, terms may therefore transcend the temporary (and temporal) objects and be considered simply as a *network of interacting processes*, which when represented in graph-theoretic form is an arrow diagram.

The Modelling Relation

5.9 Arrow Diagram When there are two systems they invite comparison.

(3)

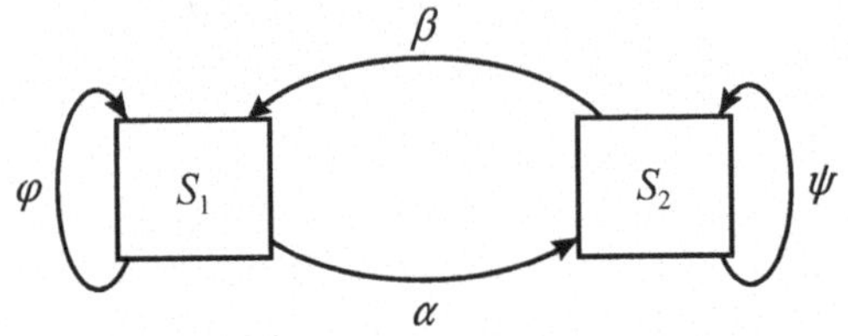

The purpose of the comparison is that one may learn something new about a system S_1 of interest by studying a different system S_2 that is its surrogate. Diagram (3) contains the components I need to describe what a *modelling relation* is between a system S_1 and a system S_2, where each system can be either natural or formal. The arrows φ and ψ represent entailment within, respectively, the systems S_1 and S_2, and as such are *intra*-system processes *internal* to their own systems.

The arrow α serves to associate features of S_1 with their counterparts in S_2, while the arrow β serves to associate features of S_2 with those of S_1. The arrows α and β taken together thus establish a kind of *dictionary*, which allows effective passage from one system to the other and vice versa. The arrows α and β are *external* to both S_1 and S_2. These *inter*-system processes are not a part of, nor are they entailed by anything in, either systems.

5.10 Simulation Since the arrows φ and ψ are internal and therefore inherent to their own systems, the vehicle for establishing a relation of any kind between S_1 and S_2 resides in the choice of the external arrows α and β. So far the relation $\langle S_1, \varphi, \alpha \rangle \leftrightarrow \langle S_2, \psi, \beta \rangle$ is symmetrical. Symmetry breaks when one system is reflected in the other.

A *necessary condition* for reflection involves all four arrows, and may be stated as 'whether one follows *path* φ or *paths* α, ψ, β *in sequence*, one reaches the same destination'. Formally, this may be expressed as the compositional equality

(4) $$\varphi = \beta \circ \psi \circ \alpha.$$

If this relation is satisfied, one says that S_2 is a *simulation* of S_1, that $\alpha : S_1 \to S_2$ is the *encoding* arrow, and $\beta : S_2 \to S_1$ is the *decoding* arrow. The reflection entails a chirality, that of an 'original' and its 'image', since evidently equation (4) is not the same as $\psi = \alpha \circ \varphi \circ \beta$; the truths of the two equality statements are independent.

Equation (4) is an abbreviation. Let f be a process in the entailment structure of the arrow φ in system S_1. It is more convenient for exposition (and without loss of generality) to represent this process, be it causal or inferential entailment, as a mapping $f : A \to B$ (whence the sets A and B represent component objects in S_1). The encoding $\alpha : S_1 \to S_2$ of systems at the component A hierarchically entails the element mapping $\alpha_A : A \to X$, where X is a component object in S_2. Consider a mapping $g : X \to Y$ (which is a process in the entailment structure of the arrow ψ, and where the sets X and Y represent component objects, in S_2) and the hierarchical action of the decoding $\beta : S_2 \to S_1$ which in turn entails the element mapping $\beta_Y : Y \to B$. Suppose the four mappings connect to make the diagram

(5)

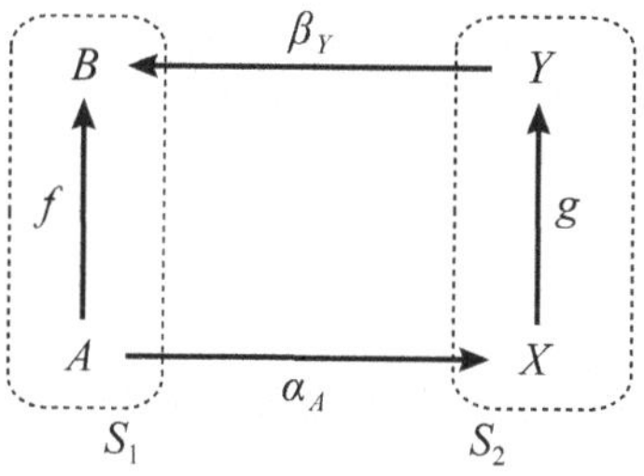

commute. This means for every element a in A, whether one traces through the mapping f alone, or through α_A followed by g and then followed by β_Y, one gets the same result in B; i.e. *for all* $a \in A$ the equality

$$f(a) = \beta_Y(g(\alpha_A(a))) \tag{6}$$

holds. Note that this commutativity condition for simulation places no further restrictions on the mapping g itself and its domain X and codomain Y, other than the relay that begins at $a \in A$, of the composite trace

$$a \mapsto \alpha_A(a) \mapsto g(\alpha_A(a)) \mapsto \beta_Y(g(\alpha_A(a))) = b, \tag{7}$$

needs to reach the correct final destination $b = f(a) \in B$. Such emphasis on the results *regardless of the manner in which they are generated* (i.e. with no particular concern on underlying principles) is the case when S_2 is a *simulation* of S_1.

5.11 Model If, however, the mapping g is *itself* entailed by the encoding α, i.e. if $g = \alpha(f)$, whence the mapping in S_2 is $a(f) : \alpha(A) \to \alpha(B)$, then one has the commutative diagram

(8)

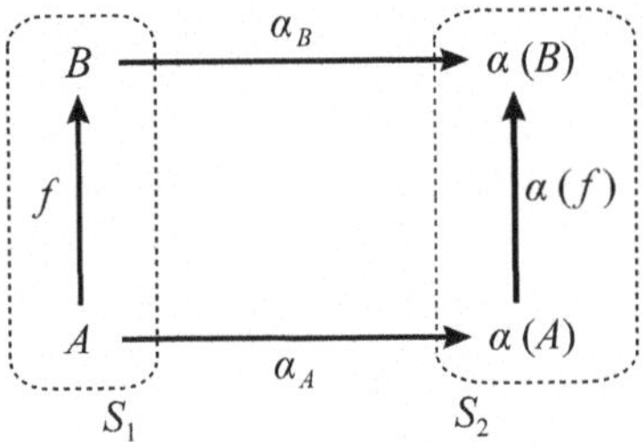

which encompasses the equality that says, for every element $a \in A$,

$$\alpha_B(f(a)) = \alpha(f)(\alpha_A(a)) \in \alpha(B). \tag{9}$$

Further, the decoding β has to suitably invert the process, so that $\alpha(f) : \alpha(A) \to \alpha(B)$ gets mapped back to a process $\beta(\alpha(f)) : \beta(\alpha(A)) \to \beta(\alpha(B))$ in S_1, such that one has the composite commutative diagram

(10)

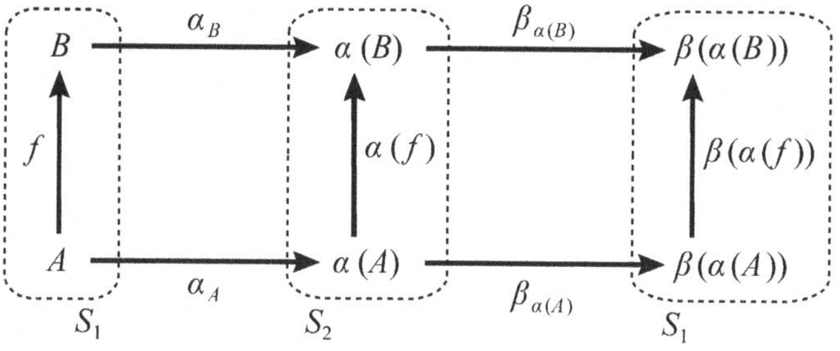

which contains, in addition to (9), the equalities

$$
\begin{aligned}
\beta_{\alpha(B)}(\alpha_B(f(a))) &= \beta_{\alpha(B)}(\alpha(f)(\alpha_A(a))) \\
&= \beta(\alpha(f))\Big(\beta_{\alpha(A)}(\alpha_A(a))\Big) \in \beta(\alpha(B)).
\end{aligned} \tag{11}
$$

Equalities (11) encompass three relays:

$$
a \mapsto \left\{ \begin{array}{l} f(a) \mapsto \alpha_B(f(a)) \mapsto \beta_{\alpha(B)}(\alpha_B(f(a))) \\ \alpha_A(a) \mapsto \alpha(f)(\alpha_A(a)) \mapsto \beta_{\alpha(B)}(\alpha(f)(\alpha_A(a))) \\ \alpha_A(a) \mapsto \beta_{\alpha(A)}(\alpha_A(a)) \mapsto \beta(\alpha(f))\Big(\beta_{\alpha(A)}(\alpha_A(a))\Big) \end{array} \right\} = b, \tag{12}
$$

and say that all three compositional paths have to reach the same element in $b \in \beta(\alpha(B))$. Further, the process $\beta(\alpha(f)) : \beta(\alpha(A)) \rightarrow \beta(\alpha(B))$ must be, in an appropriately defined sense, comparable to the original process $f : A \rightarrow B$ in S_1, such that the congruences

$$
\left\{ \begin{array}{l} \beta(\alpha(A)) \cong A \\ \beta(\alpha(B)) \cong B \\ \beta(\alpha(f)) \cong f \end{array} \right. \tag{13}
$$

hold. When these more stringent conditions are satisfied, the simulation is called a *model*. If this *modelling relation* is satisfied between the systems S_1 and S_2, one then says that there is a *congruence* between their entailment structures, that S_2 is a *model* of S_1, and that S_1 is a *realization* of S_2.

This kind of congruence between entailment structures is defined by the mathematical entity called *functor* (Definitions 0.7 and 0.9). Indeed, the commutative diagram (8) is isomorphic to the commutative diagram (38) in Section 0.17. The encoding α has a functorial representation in

$$(14)\qquad \alpha : \begin{cases} A \mapsto \alpha A & (A \in S_1 \quad \text{object}) \\ [f : A \to B] \mapsto [\alpha f : \alpha A \to \alpha B] & (f \in S_1 \quad \text{process}) \end{cases},$$

and the decoding functor β has a slightly more complicated representation in

$$(15)\qquad \beta : \begin{cases} \alpha A \mapsto \beta(\alpha A) & (\alpha A \in \alpha - \text{encoded } S_2 \text{ object}) \\ [\alpha f : \alpha A \to \alpha B] \mapsto [\beta(\alpha f) : \beta(\alpha A) \to \beta(\alpha B)] & \\ & (\alpha f \in \alpha - \text{encoded } S_2 \text{ process}) \end{cases}.$$

A simulation of a process provides an alternate description of the entailed effects. A model is a special kind of simulation that additionally also provides an alternate description of the entailment structure of the mapping representing the process itself. It is, in particular, easier to obtain a simulation than a model of a process; compare (7) with (12).

5.12 Ad Summam Simulation describes; model explains.

Natural Law

Natural order is woven into the fabric of reality. *Causality* is the principle that every effect has a cause, and is a reflection of the belief that successions of events in the world are governed by definite relations. *Natural Law* posits the existence of these *entailment* relations *and* that this causal order can be *imaged* by implicative order (*ML*: 4.7).

5.13 The Canon A *modelling relation* is a commutative functorial encoding and decoding from one system to another. (Note that the chirality inherent in the symmetry-breaking reflection is still in place.) From a natural system N to a formal system M, the situation may be represented in the following canonical diagram:

(16) 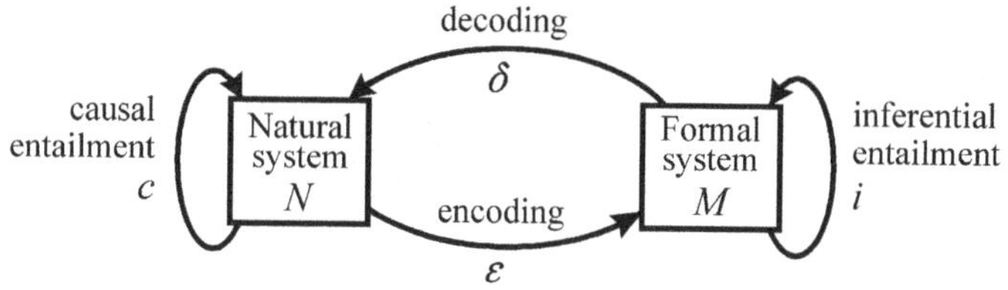

(I have replaced the symbols $\langle S_1, \varphi, \alpha \rangle \to \langle S_2, \psi, \beta \rangle$ in the generic diagram (3) with the specific $\langle N, c, \varepsilon \rangle \to \langle M, i, \delta \rangle$.) The encoding ε maps the natural system N and its causal entailment c therein to the model formal system M and its internal inferential entailment i; i.e., ε acts functorially on objects and processes thus:

(17)
$$\varepsilon : \begin{cases} N \to M \\ c \to i \end{cases}.$$

The decoding δ does the reverse on the encoded objects and processes, mapping the formal system $\varepsilon(N) \subset M$ to its realization $\delta(\varepsilon(N) \subset N)$:

(18)
$$\delta : \begin{cases} \varepsilon(N) \to \delta(\varepsilon(N)) \\ \varepsilon(c) \to \delta(\varepsilon(c)) \end{cases}.$$

The commutativity condition (4), when translated into these initial-predicated notations, becomes

(19)
$$c = \delta \circ i \circ \varepsilon,$$

which says that tracing through the *c*ausal entailment arrow c is the same as tracing successively through the three arrows, εncoding ε, *i*nferential entailment i, and δecoding δ.

5.14 The Axioms A model formal system may simply be considered as a *set* with additional mathematical structures, i.e., a concrete category (Definition 0.16). So the mathematical statement $\varepsilon : N \to M$, i.e., the immanent causation (cf. *ML*: 5.18) that posits the existence for every natural system N a model formal system M, may be stated as the axiom

(20) Everything is a set.

Causal entailment in a natural system is a network of interacting processes, i.e., a network of efficient causes. The mathematical statement $\varepsilon : c \to i$, i.e., the functorial correspondence of morphisms, between causality c in the natural domain and inference i in the formal domain, may thus be stated as an epistemological principle, the axiom

(21) Every process is a mapping.

Together, the two axioms (20) and (21) are the mathematical foundation of Natural Law. These self-evident truths serve to explain "the unreasonable effectiveness of mathematics in the natural sciences". In adopting these two axioms, we, in the Rashevsky-Rosen school of relational biology, take implicitly as the mathematical foundation of our science concrete categories. Explicitly, we shall be considering *non-full subcategories* **C** of **Set**, in which **C**-objects are sets $A, B, \ldots$, and **C**-hom-sets $H(A, B)$ are proper subsets of $\mathbf{Set}(A, B) = B^A$. By extension, we may also consider concrete categories **C**, equipped with faithful functors from **C** to non-full subcategories of **Set**, for which **C**-objects are sets with structures and **C**-hom-sets contain morphisms that preserve these structures.

5.15 Mathematical Formulation of Natural Law The collection of all models of a natural system N is denoted $\mathbf{C}(N)$ (*ML*: 7.27). $\mathbf{C}(N)$ is a lattice (*ML*: 7.28) as well as a category (*ML*: 7.29). Since each model formal system $M \in \mathbf{C}(N)$ is itself a category, one sees that $\mathbf{C}(N)$ may be considered a subcategory of **Cat** (cf. Section 0.13). Let $\kappa(N)$ be the collection of all efficient causes in N (*RL*: 7.10). An entailment network that models N may be denoted $\varepsilon(N) \in \mathbf{C}(N)$, and the collection of all efficient causes in the network $\varepsilon(N)$ may be represented by the encoding $\varepsilon(\kappa(N))$. The model formal system M of N is thence $\langle \varepsilon(N), \varepsilon(\kappa(N)) \rangle$, i.e., M is the category **C** with $\mathfrak{O}\mathbf{C} = \varepsilon(N)$ and $\mathfrak{A}\mathbf{C} = \varepsilon(\kappa(N))$. Natural Law is the statement

$$
\begin{aligned}
&\forall N\, \exists \varepsilon\, \exists M \in \mathbf{C}(N) : M = \varepsilon(N) \\
&\wedge\ \forall c \in \kappa(N)\, \exists i \in \varepsilon(\kappa(N)) : i = \varepsilon(c).
\end{aligned}
\tag{22}
$$

For notational simplicity, however, one often drops the encoding symbol ε and uses N to denote both the natural system and its network model that is a formal system. Thus 'an entailment network $\varepsilon(N)$ that models a natural system N' abbreviates to 'an entailment network N'. Likewise the symbol $\kappa(N)$ shall denote the collection of efficient causes in both the natural system and the formal system.

The relational diagram of $\kappa(N)$ is a digraph representing the entailment network N (*RL*: Chapter 9). These identifications $\varepsilon(N) = N$ and $\varepsilon(\kappa(N)) = \kappa(N)$, whence $\mathfrak{O}\mathbf{C} = N$ and $\mathfrak{A}\mathbf{C} = \kappa(N)$, amount to an implicit invocation of Natural Law. The existence of causal entailment in a natural system is ontological; the representation of causality, by an arrow (i.e., as mappings), is epistemological.

Every Process is a Set-Valued Mapping

5.16 Towards Plurality Axiom (21) carries with it the baggage inherited from Newtonian physics, which mathematically is the analysis of single-valued mappings. In view of the sparsity (Section 2.6) of single-valued mappings among set-valued mappings, I now rephrase axiom (21) as

(23) Every process is a set-valued mapping.

Stated otherwise, I am expanding our theatre from the category **Set** to the category **Rel**. Because of the containment $\mathbf{Set}(X, Y) \subset \mathbf{Rel}(X, Y)$, however, a set-valued mapping may incidentally be single-valued, so the new extension encompasses the old foundation. A model formal system M of a natural system N is now a subcategory of **Rel**, and by extension a 'concrete' category $\mathbf{C}$ equipped with a faithful functor $F : \mathbf{C} \to \mathbf{Rel}$.

The extension of the Natural Law axiom from (21) to (23) realizes the fact that processes inevitably entail more than their single, primary, 'intended' outputs. When the extraneous secondary outputs are material causes, they are called *by-products*; when they are efficient causes, they are called *side-effects*. Although these two terms often have negative connotations, and are predominantly employed to describe adverse outputs, they can as well apply to beneficial, albeit still unanticipated, consequences. In the next chapter I shall illustrate with a few examples.

5.17 … ob der liebe Gott würfelt It is important to note my theory of 'set-valuedness' of process outputs has nothing to do with fuzziness or probabilistic reasoning. «Je n'avais pas besoin de cette hypothèse-là.» There is no difference in determinacy between set-valued mappings and their specialized single-valued comrades, neither necessitating stochastics in their formulations. Axiom (23), 'Every process is a set-valued mapping.', has to do with the well-connectedness of the entailment networks of Nature, that every action entails a plenitude of consequences. Think 'butterfly effect': chaotic systems are sensitive to initial conditions, but are still deterministic.

5.18 E pluribus unum The teleology of "One. True. Purpose." of a natural process is an artefact. A process simply is; its plural outputs simply are. Members of an output set (i.e., constituents of a 'value' of a set-valued mapping) are coextensive. All those materially entailed are products; all those functionally entailed are effects. The existence of a meta-process of value judgment is an implicit bias, when one distinguishes a 'desired' product and adds the prefix 'by-' to the rest, and likewise when one isolates an 'intended' effect and relegates others to be qualified with 'side-'.

Given a set-valued mapping $F : X \multimap Y$, the teleological assignment of a single final cause is a *choice mapping*

$$\varepsilon : \{F(x) \subset Y : x \in X\} \to Y \tag{24}$$

that selects, for each $x \in X$, from the set $F(x) \subset Y$ a single value

$$\varepsilon(F(x)) \in F(x) \subset Y; \tag{25}$$

the procedure thus defines a single-valued mapping

$$f = \varepsilon \circ F : X \to Y. \tag{26}$$

(See *RL*: 0.20 for the definition of choice mapping, and *RL*: 9.3 for its role in an explication of functional closure.) The use of the same symbol ε for the encoding functor and the choice mapping is not coincidental: the single-purpose specialization

$$\hat{\varepsilon} : F \mapsto f \tag{27}$$

is the encoding

$$\hat{\varepsilon} : \kappa(N) \to \hat{\varepsilon}(\kappa(N)) \tag{28}$$

in the modelling of a natural system N with set-valued processes by its abstraction with single-valued processes.

As with any model, the commitment to a specific choice mapping ε, whence a particular encoding $\hat{\varepsilon}$ projecting set-valued mappings F to single-valued mappings f, loses information in its execution. In this case, the loss of 'degrees of freedom', 'closing' each set $F(x)$ in the restriction to a single output $\varepsilon(F(x)) \in F(x)$, renders unentailed all the by-products and side-effects, which consequently become unexplainable in the decoding. To wit, the encoding is a simple choice (a possible invocation of the Axiom of Choice 2.22 notwithstanding) of a single element $y_x = \varepsilon(F(x)) \in F(x)$ from each member of a family $\{F(x) : x \in X\}$ of subsets of Y; but there is no trivial procedure for the from-one-to-all extrapolation $\delta : \{y_x : x \in X\} \to \mathrm{P}Y$. The decoding, from a paucity of ingredients, has the futile task to reconstruct from each member y_x of the indexed set $\{y_x : x \in X\}$ to a superset $\delta(y_x)$, such that $y_x \in \delta(y_x) \subset Y$ and the reconstitution $F(x) = \delta(y_x)$ recovers the original set-valued mapping $F : X -\!\sqsubset Y$. This loss of entailment closes functionally open systems, and this informational incompleteness is, indeed, what Robert Rosen proposed as the cause of side-effects [Rosen 1985]. I shall discuss this further in the next chapter.

5.19 Ut sint unum sicut et nos Throughout *RL* and heretofore in *IL*, I have almost always used capital letters to denote set-valued mappings (e.g., F in $F : X -\!\sqsubset Y$) and lowercase to denote standard (single-valued) mappings (e.g., f in $(f : X \to Y)$. The distinction has been useful, for the sake of clarity, when comparing, contrasting, and connecting the two kinds of morphisms. But in fact the two species of arrows, $-\!\sqsubset$ and $\to$, are sufficient in themselves to distinguish the formal causes (cf. Section 3.1).

Mathematicians often run out of symbols in their expositions even with at their disposal both the Greeks alphabet and the Latin alphabet (and the latter rendered in various fonts besides). Henceforth, especially in view of the plurality axiom (23), I shall use lowercase letters to denote set-valued mappings (which may incidentally be single-valued) that are general processes, e.g. $f : X -\!\sqsubset Y$. This way, I have freed up capital letters to represent other entities.

The Many Levels of the Encoding Functor

True to their category-theoretic taxonomy as *functors*, the encoding ε and the decoding δ map on many levels, succinctly manifested in the commutative diagram:

(29)
$$\langle N, \kappa(N) \rangle \underset{\varepsilon}{\overset{\delta}{\rightleftarrows}} \langle M, \kappa(M) \rangle.$$

5.20 Model Level On the category-of-models level,

(30)
$$\varepsilon : N \to \mathbf{C}(N),$$

The encoding functor ε assigns to each representation $\langle N, \kappa(N)\rangle$ of N a model system

(31) $$\langle M, \kappa(M)\rangle = \langle \varepsilon(N), \varepsilon(\kappa(N))\rangle$$

in $\mathbf{C}(N)$. The assignment $\varepsilon : N \mapsto M$ is another choice mapping; it singly selects, as a specific model of the natural system N, the formal system M from the set $\mathbf{C}(N)$.

In view of the Natural Law statement (22) and the identification made in Section 5.15 above, encoding ε entails, for each set $A \in N = \mathfrak{O}\mathbf{C}$ and for each mapping $f \in \mathbf{C}(A,B) \subset \kappa(N) = \mathfrak{A}\mathbf{C}$, that

(32) $$\varepsilon : \begin{cases} A \mapsto \varepsilon(A) \in \varepsilon(N) = M \subset \mathfrak{O}\mathbf{C} & (A \in \mathfrak{O}\mathbf{C}) \\ f \mapsto \varepsilon(f) \in \mathbf{C}(\varepsilon(A), \varepsilon(B)) \subset \kappa(M) \subset \mathfrak{A}\mathbf{C} & (f \in \mathfrak{A}\mathbf{C}) \end{cases}.$$

An *ideal* (but almost never achieved) decoding would map functorially thus:

(33) $$\delta : \begin{cases} X \mapsto \delta(X) = A \subset N = \mathfrak{O}\mathbf{C} & (X = \varepsilon(A) \in M \subset \mathfrak{O}\mathbf{C}) \\ g \mapsto \delta(g) = f \in \kappa(N) = \mathfrak{A}\mathbf{C} & (g = \varepsilon(f) \in \kappa(M) \subset \mathfrak{A}\mathbf{C}) \end{cases}.$$

How close one can make the decoding functor δ behave as in (33) is the subject of Part III, later here in *IL*.

5.21 Point Level In addition to the *set-pairing* $(N, M) \in \varepsilon$, ε also functions on the *point-pairing* level as a mapping $\varepsilon_A : A \to \varepsilon(A)$ for each $A \in N = \mathfrak{O}\mathbf{C}$ *from one set into another*. To each input element (material cause) $n \in A \in N$, there corresponds a unique output element (final cause) $m \in M$ such that $(n, m) \in \varepsilon_A$; i.e., $\varepsilon_A : n \mapsto m$. The component mappings may be collected into the family

(34) $$\varepsilon = \{\varepsilon_A : A \in N = \mathfrak{O}\mathbf{C}\}.$$

which is abbreviated into the point-set map notation

(35) $$\varepsilon : N \to M.$$

5.22 Process Level The trace $\varepsilon : c \mapsto i$ is a functorial correspondence of morphisms

(36) $$\varepsilon : \kappa(N) \rightarrow \kappa(M).$$

This *process-pairing* $(c, i) \in \varepsilon$ functions on a higher hierarchical level than point-pairing, because now the output is *itself* a mapping $i = \varepsilon(c) \in \kappa(M)$.

5.23 Commutative Diagram The commutativity condition (19) may be drawn as the element trace

(37)

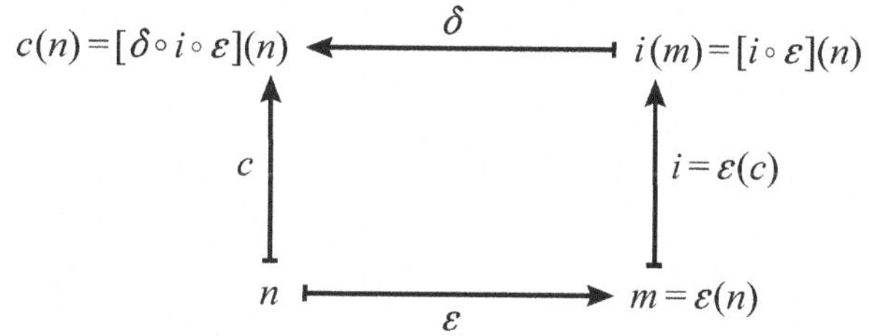

The encoding functor ε thus encompasses both kinds of entailment in its effects: the output $m = \varepsilon(n) \in M$ is a point, and $\vdash m$ is material entailment; the output $i = \varepsilon(c) \in \kappa(M)$ is a mapping, and $\vdash i$ is functional entailment.

5.24 Category of Diagrams Recall (Section 0.21) that the functor category $\mathbf{D}^{\mathbf{C}}$ may be considered to contain copies of all diagram of **C**-morphisms that may be formed from the objects and morphisms of **D**. The transition from a category **C** to a functor category $\mathbf{D}^{\mathbf{C}}$, then, creates *models* of subcategories of **C** in **D**. In particular, the encoding functor $\varepsilon : N \rightarrow \mathbf{C}(N)$ is an object in the functor category $\mathbf{C}(N)^N$.

Note the phrase 'that may be formed from the objects and morphisms of **D**' in the category-of-diagrams characterization of $\mathbf{D}^{\mathbf{C}}$: the availability of models of N depends on the richness of the codomain category $\mathbf{D} = \mathbf{C}(N)$. The essence is that the concept of natural transformation involves the existence of suitable objects and morphisms in the category $\mathbf{C}(N)$ of models that would render the appropriate diagrams commutative. Thus, the modelling relation must always be defined relative to a particular category, and not in absolute terms. The subjective choice of which subcategory $\mathbf{C}(N)$ of **Cat** to use as one's universe of models is, then, the art of the craft of modelling.

We have already encountered the relational-biologic realization of this category-theoretic concept. In *RL*: 8.1, I concluded that in a suitably equipped category (in the sense of an appropriate selection of morphisms), any sufficiently large finite family of morphisms must inevitably contain a mutual connection that is *functional entailment*. This culminated in the important

5.25 Postulate of Biopoiesis (*RL*: 8.29) In a suitably equipped category, any sufficiently large finite family of morphisms must contain an (M,R)-system.

6
Opera
By-Products and Side-Effects

— Orlande de Lassus
[Orlando di Lasso] (1532–1594)
Cantabant canticum Moysi

A. H. Louie, *Intangible Life*, Anticipation Science 2,
DOI 10.1007/978-3-319-65409-6_7

It is now opportune to illustrate the category theory of set-valued mappings with some biological implications. I shall explicate another aspect of relational pathophysiology (a subject that I first investigated in Louie [2012] and then *RL*: Chapters 12–14): the general problems connected with system error, malfunction, and breakdown — by-products and side-effects.

Sequential Composites

An alternative to the sequential composite may be defined for set-valued mappings.

6.1 Square Product Let $f : X \multimap Y$ and $g : Y \multimap Z$ be set-valued mappings. Their *square product* is the set-valued mapping $g \,\square\, f : X \multimap Z$ defined by, for $x \in X$,

(1) $$(g \,\square\, f)(x) = \bigcap_{y \in f(x)} g(y) \subset Z.$$

Note that *the codomain of f is the domain of g*, enabling the composition that is the sequence '*f* followed by g'.

Recall (Definition 3.5) the *sequential composite*, the set-valued mapping $g \circ f : X \multimap Z$ defined by, for $x \in X$,

(2) $$(g \circ f)(x) = \bigcup_{y \in f(x)} g(y) \subset Z,$$

that is the composite operation in the category **Svm** (Definition 3.7). Note the different symbols used for the two binary operations: for sequential composition in (2) it is the standard 'small circle' $\circ$ of 'composite'; for square product in (1) it is a 'small square' $\square$. Recall (Section 3.11) that the graph of the sequential composite of set-valued mappings is the relative product of their graphs. Under the **Rel** $\cong$ **Svm** isomorphism, the sequential composite is also the composite operation in the category **Rel**.

The same steps in the demonstration (in Section 3.5) of the associativity of the sequential composite may be used (with the trivial replacement of $\cup$ by $\cap$) to verify that the square product is associative. Thus, for set-valued mappings $e : W \multimap X, f : X \multimap Y, g : Y \multimap Z$, and $w \in W$,

(3) $$(g \,\square\, f) \,\square\, e = g \,\square\, (f \,\square\, e) : W \multimap Z.$$

6.2 Paths As relations, subsets of $X \times Z$, the two sequential composites (1) and (2) become

(4) $$(g \,\square\, f) = \{(x,z) \in X \times Z : \forall y \in Y \quad (x,y) \in f \Rightarrow (y,z) \in g\}$$

and

(5) $$(g \circ f) = \{(x,z) \in X \times Z : \exists y \in Y \quad (x,y) \in f \wedge (y,z) \in g\}.$$

From the $\exists$ – versus – $\forall$ characterization, one may consider that, for each element $x \in X$ as it is mapped by f into Y and then by g into Z, the sequential composition $(g \circ f)(x)$ traces *at least one* path while the square product $(g \,\square\, f)(x)$ traces *all* such paths. In either case, when one has $z \in (g \circ f)(x)$ or $z \in (g \,\square\, f)(x)$ (equivalently, when one has $(x,z) \in (g \circ f)$ or $(x,z) \in (g \,\square\, f)$), the implication is that, for at least one element $y \in f(x)$, the element-trace relay $x \mapsto y \mapsto z$ is possible.

6.3 Relations between $\circ$ and $\square$ Since $A \cap B \subset A \cup B$ and '$\forall \prec \exists$', one may be tempted to conclude from the definitions of the two species of sequential composites that $(g \,\square\, f)(x) \subset (g \circ f)(x)$. But the situation is more subtle, and it depends on whether $|f(x)| > 1$, $= 1$, or $= 0$.

The containment $(g \,\square\, f)\,(x) \subset (g \circ f)(x)$ is indeed true when $|f(x)| \geq 1$. When $|f(x)| > 1$, the containment is often proper. It may incidentally happen that for $a, b \in f(x)$, $g(a) \cup g(b) \neq \varnothing$ whence $(g \circ f)(x) \neq \varnothing$, but $g(a) \cap g(b) = \varnothing$ whence $(g \,\square\, f)(x) = \varnothing$. In such a case, there are paths to relay $x \mapsto f(x) \mapsto (g \circ f)(x)$, but there is no common destination for all paths beginning with x and processed sequentially by f then g. One therefore sees that the square product $(g \,\square\, f)(x)$ is more stringent in its relay than the sequential composite $(g \circ f)(x)$.

When $|f(x)| = 1$, the intersection in (1) and the union in (2) are both taken over the same single element, hence $(g \,\square\, f)(x) = (g \circ f)(x)$. For standard (single-valued) mappings, for each $x \in X$ $|f(x)| = 1$, the two sequential composites (1) and (2) therefore coincide (and are identical to the standard sequential composition of mappings).

When $|f(x)| = 0, f(x) = \varnothing$. One has to be aware of 'empty set pathologies' from taking the union and intersection over an empty set. In a lattice L, the least element and greatest element (when they exist) are $\inf L = \sup \varnothing$ and $\sup L = \inf \varnothing$. For the lattice $\mathrm{P}Z$, one has $\varnothing = \inf \mathrm{P}Z = \bigcup_{\varnothing}$ and $Z = \sup \mathrm{P}Z = \bigcap_{\varnothing}$ (cf. *ML*: 1.28). Therefore, in the singular scenario $f : x \mapsto \varnothing$, when f acting on input x produces no outputs, $(g \circ f)(x) = \bigcup_{y \in \varnothing} g(y) = \varnothing$ and $(g \,\square\, f)(x) = \bigcap_{y \in \varnothing} g(y) = Z$, which a fortiori says $(g \,\square\, f)(x) \not\subset (g \circ f)(x)$. One may explain the apparent paradox thus. The existence of $z \in (g \circ f)(x)$ implies the existence of an element $y \in f(x)$, for which $z \in g(y)$, to complete the relay path $x \mapsto y \mapsto z$. But $f(x) = \varnothing$ means no such relay point $y \in f(x)$ can exist, so contrapositively there can be no $z \in (g \circ f)(x)$; whence $(g \circ f)(x) = \bigcup_{y \in \varnothing} g(y) = \varnothing$. On the other hand, for an element $z \in Z$ to satisfy $z \in (g \,\square\, f)(x)$, it must happen that whenever $y \in f(x)$, the relay $x \mapsto y \mapsto z$ ensues. But $f(x) = \varnothing$ means the implication '$y \in f(x) \Rightarrow$ relay $x \mapsto y \mapsto z$' is vacuously true (i.e., there is no $y \in f(x)$ to contradict the statement), so every $z \in Z$ qualifies; whence $(g \,\square\, f)(x) = \bigcap_{y \in \varnothing} g(y) = Z$.

6.4 Relational Diagrams In Chapter 3 of *RL*, I have used the same symbology (Notations 2.2) of $\langle \text{solid-headed arrow} + \text{hollow-headed arrow} \rangle$ pair for the relational diagram of a set-valued mapping. This latter formal cause, however, did not make much of a reappearance after its debut. I now, in anticipation of the exploration to come, think it is appropriate to introduce variations on the theme.

Instead of the 'hollow-*triangle*-headed arrow' of a single-valued mapping, I propose new formal causes in both a 'hollow-*circle*-headed arrow' and a 'hollow-*square*-headed arrow' for 'that which is entailed' in the set-valued mapping $f : x \mapsto B$ (where $x \in X$ and $B = f(x) \subset Y$):

(6)

(7)

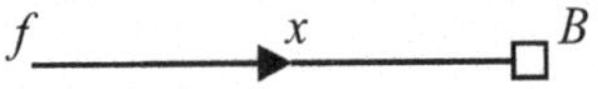

Both the circle-headed and square-headed species serve to emphasize that the final cause (output) is a set, and the two species indicate the two different kinds of compositions that may be involved.

The relational diagram of sequential composition $(g \circ f)(x)$ is

(8)

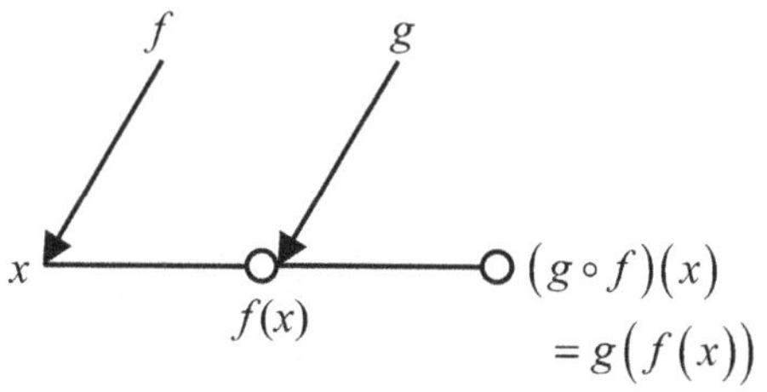

The crucial node in the entailment network (8) is the 'relay vertex' $f(x)$, where the two set-valued mappings $f : X -\!\sqsubset Y$ and $g : Y -\!\sqsubset Z$ interact and their relational diagrams connect. That $f(x)$ is where its hollow-circle-headed arrow and the solid-headed arrow of g meet represents the execution of the *union* operation

(9)
$$\bigcup_{y \in f(x)} g(y),$$

which is the very definition of $(g \circ f)(x)$. The iconography of this connection, stated otherwise, symbolizes a sequential relay of *at least one* output in $f(x)$ as a material cause of g; at least one elemental path must pass through here to reach $(g \circ f)(x)$. The hollow-circle-headed arrow that terminates on $(g \circ f)(x) = g(f(x))$ simply indicates that the final output is a set, but it could have been replaced by a hollow-square-headed arrow. It is the further relay of this output that would determine the alternatives of circle and square.

The relational diagram of square product $(g \,\square\, f)(x)$ is

(10)

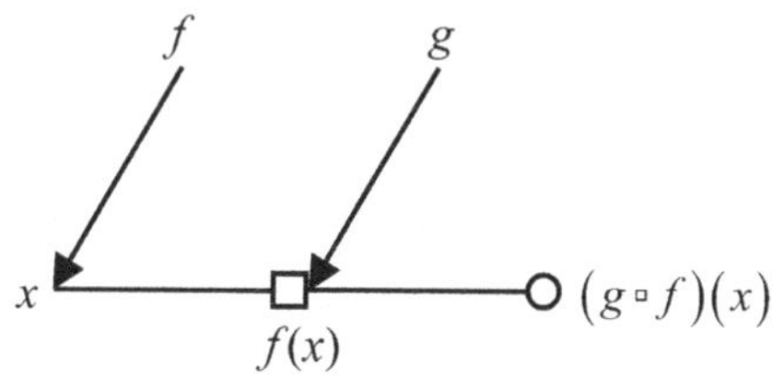

Here the node $f(x)$ is where its hollow-square-headed arrow and the solid-headed arrow of g meet; the connection represents the execution of the *intersection* operation

$$(11) \qquad \bigcap_{y \in f(x)} g(y),$$

which is the very definition of $(g \,\square\, f)(x)$. This iconography, then, symbolizes a sequential relay of *all* the outputs in $f(x)$ as a material causes of g; all elemental traces through the relay vertex $f(x)$ must terminate at the final output $(g \,\square\, f)(x)$. As before, the hollow-circle-headed arrow that terminates on the final output simply indicates that the latter is a set, with a possibility of a symbol-change if this output is further relayed.

By-Products

> Happiness is not a goal, it is a by-product. Paradoxically, the one sure way not to be happy is deliberately to map out a way of life in which one would please oneself completely and exclusively.
>
> — Eleanor Roosevelt (1960)
> *You Learn by Living*

6.5 Respiration Cellular respiration is the metabolic process that takes place in the cells of organisms to convert biochemical energy from nutrients into a useful form to fuel cellular activities. It may be succinctly represented by this relational diagram:

(12)

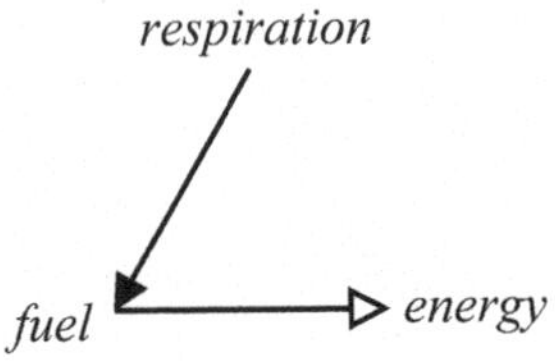

The reductionistic study of this process (or set of processes) forms the bulk of the curriculum of biochemistry. There is no need here to get into the intricacies of glycolysis, citric acid cycle, Krebs cycle, etc. It suffices to state (aerobic) respiration in its simplest form:

(13) $$C_6H_{12}O_6 + 6\,O_2 \rightarrow 6\,CO_2 + 6\,H_2O\ +\ \textit{energy}.$$

Water is useful (in moderation) in the cellular environment. Carbon dioxide is a toxin, and higher organisms have elaborate respiratory systems for its removal. Both H_2O and CO_2 are by-products for the *raison d'être* of respiration that is useful-energy generation. The relational diagram (12) is thus more accurately represented as a set-valued mapping:

(14)

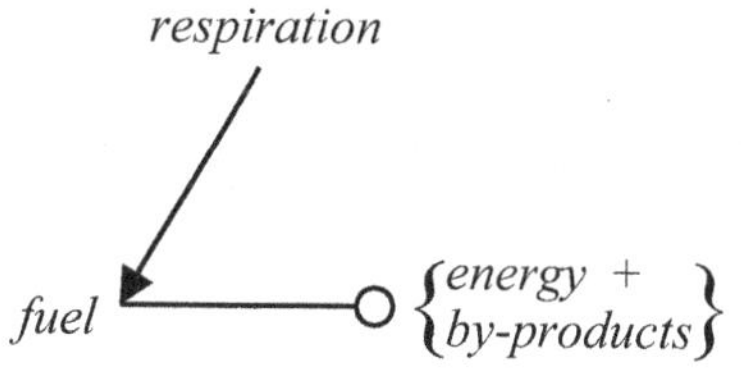

6.6 Photosynthesis Photosynthesis is a process used by plants (and other organisms) to convert light energy into chemical energy that can be used to fuel the organisms' activities. Biochemistry students also study, of course, photosynthesis in terms of the biochemical pathways of the Calvin cycle. The simplest chemical equation for oxygenic photosynthesis is

(15) $$2n\,CO_2 + 2n\,H_2O + \textit{light energy} \rightarrow (C\,H_2O)_{2n} + 2n\,O_2.$$

Photosynthetic organisms are photoautotrophs: they are able to synthesize food directly from carbon dioxide and water using light energy. Oxygen is released, mostly as a waste product (of the photosynthetic process itself; oxygen is otherwise crucial in the biosphere for all the oxygen-breathing organisms). In effect, photosynthesis is the 'inverse process' of cellular respiration, and its set-valued mapping has this relational diagram:

(16)

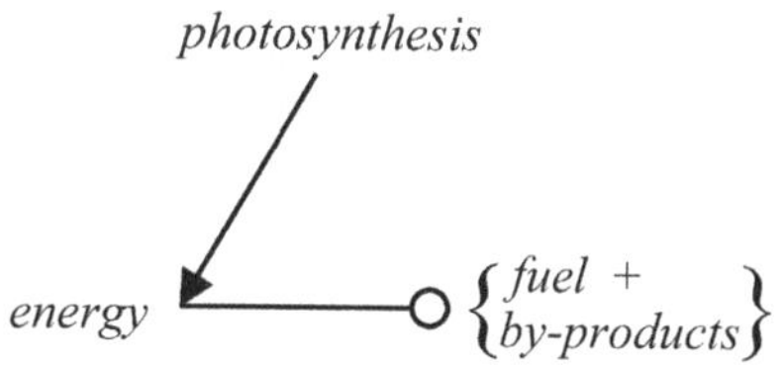

6.7 Teleological Inverses It is a testament to the wonder of Nature that two complimentary processes, (14) and (16), exist in the biosphere to precisely use each other's by-products as input material causes. One must note, however, that the two processes take place through a different sequence of chemical reactions and in different cellular organelles and compartments. While photosynthesis and cellular respiration are mutual teleological inverses in terms of entailment, they do not 'reverse' each other's efficient causes. In short, as set-valued mappings,

$$\text{(17)} \qquad \textit{photosynthesis} \neq (\textit{respiration})^{-1}.$$

Other examples of (positive and negative) by-products include: manure from animal husbandry, straw from grain harvesting, the plethora of 'petroleum products' from the refining of crude oil to produce gasoline, radioactive waste from nuclear power generation, 'green house gases' from combustion. The reader may easily append to this list.

Side-Effects

> I think of my body as a side-effect of my mind. Like a thought I had once that manifested itself — Oops! Oh no! Manifested. Look at this. Now we have to buy clothes and everything.
>
> — attributed to Carrie Fisher

6.8 Unanticipated Consequences The term 'side-effect' is borrowed from medicine, and describes unavoidable and usually unfortunate consequences of employing therapeutic agents. Robert Rosen has considered the subject in Rosen [1974a, b] from the viewpoint of 'how planning could go wrong', and devoted a substantial portion of the introductory Section 1.1 of *Anticipatory Systems* [Rosen 1985] on a repeat discussion. In the latter, a control-theoretic definition of 'side-effect' is found:

> unplanned and unforeseeable consequences on system behavior arising from the implementation of controls designed to accomplish other purposes; or, in a related context, the appearance of unpredicted behavior in a system built in accordance with a particular plan or blueprint

followed by Rosen's explication:

> … the ultimate seat of the side-effects arising in anticipatory control, and indeed of the entire concept of error or malfunction in system theory as a whole, rests on the discrepancy between the behavior actually exhibited by a natural system, and the corresponding behavior predicted on the basis of a model of that system. For a model is necessarily an *abstraction*, in that degrees of freedom which are present in the system are absent in the model. In physical terms, the system is *open* to interactions through these degrees of freedom, while the model is necessarily *closed* to such interactions; the discrepancy between system behavior and model behavior is thus a manifestation of the difference between a closed system and an open one.

Thus Rosen contends that side-effects are necessarily entailed from the exercise of control. Note the adjectives 'unplanned', 'unforeseeable', and 'unpredicted' in Rosen's definition are all *model-dependent* qualifiers. The categorization of certain outputs as '*un*anticipated consequences', as *side*-effects (and likewise as *by*-products), is an artificial value-judgment, a subjective imposition of 'purpose', an extrinsic choice in model-making (cf. my explication in Section 5.18 above).

6.9 Polysemous Entailment Rosen's epistemological explanation of the exodus of side-effects, as perused from Rosen [1974a, b], is that it is a natural consequence of every model abstraction, entailed from 'something relevant has been missed'. (And, depending on the quality of the chosen predictive models, some models miss more than others.) I am offering a complementary, ontological explanation, that the genesis of side-effects is a natural consequence of the multitudinous outputs of every process, entailed from 'something extraneous is concurring'. Both explanations are, nevertheless, ultimately anchored on the same premise that natural processes multitask, that they entail polysemously, that, irrevocably, 'Every process is a *set-valued* mapping.' (cf. Section 5.16)

The ideal of the *magic bullet*, a substance or a therapy capable of providing a remedy for an ailment without deleterious side effects, is not in general achievable. It is also unlikely that side-effects can be removed by simply augmenting the model, or by attempting to control each side-effect separately as it appears. (Think of the stereotypical cartoon of the Little Dutch Boy who stuck a finger into a leaking hole in a dike to stem the flow of water, only to have another leak sprung up elsewhere, and eventually ran out of additional digits to plug an ever-increasing number of holes.) This is because generic natural systems are complex (in the Rosen sense; cf. *ML*: Chapter 9). Only simple models have simulable largest models, which, when used, produce no side-effects, because thence every entailed effect has been accounted for. Indeed, all aspects of simple system behaviour, from generation to revelation, are explicitly embodied in the finitely generated syntax of that largest model. At times it is possible to avoid the divergence of an infinite sequence of patchwork fixes in complex system modelling; this is when certain types of functional closure in the system can be implemented. Closure to efficient causation and its corollaries are not, however, the subject herewith; these have been addressed in *ML* and *RL*.

6.10 The Good Intentions Paving Company Saint Bernard of Clairvaux wrote (*c*. 1150) «L'enfer est plein de bonnes volontés et désirs.» Side-effects are the curse of chemical therapeutics. There is no need to dwell herein on this well-known failure, except to note that almost all therapeutic and diagnostic agents have side-effects. A horrific example, to give but one for illustration, of such paved road to hell is the thalidomide disaster in the late 1950s and early 1960s. Occasionally, some side-effects are, however, beneficial. For instance, acetylsalicylic acid (ASA, "Aspirin"), which is a pain reliever primarily (which really only signifies a historical accident that its analgesic properties were discovered first), has a secondary unintended effect as an anticoagulant that can help prevent heart attacks and reduce the severity and damage from thrombotic strokes. A beneficial side-effect, once discovered, may lead to the drug's function change, so that the secondary effect becomes one of the primary ones. ASA is now widely prescribed in its preventive role.

The introduction of a non-native species into an ecosystem for an intended purpose (pest control, decoration, recreation, ...) often 'upsets the balance of nature' (when the introduced species flourishes due to the lack of natural predators in the new environment), and does more harm than good. Think of rabbits in Australia, and Africanized bees in the Americas. Many evolutionary changes in organisms are unintended consequences, side-effects of otherwise-purposed processes. This is the *principle of function change*, and Rosen's favourite example [Rosen 1974a] is how the swim bladders of fishes (intended as an organ of equilibration, a 'flotation device') evolved into lungs for air-breathers.

Sometimes side-effects are unanticipated because of the fractionation of a complex system into its analytic parts (cf. *ML*: 8.31–8.33): a side-effect in the whole integrated system may be abstracted away in a fractionated subsystem. Side-effects often only reveal themselves after some time delay. The atomic bomb brought World War II to a quick end, but the accompanying costs—of radioactive fallout, electromagnetic pulse, nuclear winter, and so on—were not discovered until later. Dichlorodiphenyltrichloroethane (DDT) kills insects, and the concentration when used as insecticide is arguably mostly harmless. But the complexity of an ecosystem is intricately integrated. Laboratory and field tests had not anticipated the emergent biological phenomenon known as *biomagnification*, when the concentration of a substance in an organism exceeds that in the background, and the concentration increases as the substance moves up the food chain. By the time biomagnification was offered as an explanation after birds began to disappear, extreme high levels of toxic DDT were found in human breast milk. DDT is now a banned substance in many parts of the world.

Many social and economic policies introduced with good intentions have not only generated unfortunate side-effects, but have actually, in the long run, served to exacerbate the very problems they were meant to control. Examples of such perverse functional entailments abound: prohibition entailed alcohol-smuggling organized crime, highways without curves induced road hypnosis, an attempt to censor or remove a certain piece of information (document, photograph, video, etc.) from the internet instead caused the information to 'go viral' and become widely known and distributed. The reader is invited to supply more examples of positive and negative side-effects.

The Imminence Mapping

In *RL*, the theory of set-valued mappings culminates in the subtitular subject of *imminence mapping*, which equips the further investigation of functional entailment in complex relational networks. Imminence in (M,R)-networks that model living systems addresses the topics of biogenesis and natural selection. Interacting (M,R)-networks with mutually entailing processes serve as models in the study of symbiosis and pathophysiology. The formalism also provides a natural framework for a relational theory of virology and oncology.

Let us see what the imminence mapping may entail on the subject of by-products and side-effects.

6.11 Models in $\mathbf{C}(N)$ Let N be a natural system. On our path towards plurality (Section 5.16), I have defined a model formal system M of a natural system N to be a subcategory of **Rel** (and by extension a concrete category **C** equipped with a faithful functor $F : \mathbf{C} \to \mathbf{Rel}$).

In the category **Rel**, without restriction, models of N are built on **Rel**-objects (sets) and **Rel**-morphisms (set-valued mappings). In general our models are drawn from smaller non-full subcategories **C** of **Rel**, in which **C**-objects are a selection of sets $A, B, \ldots$, and **C**-hom-sets are proper subsets of $\mathbf{Rel}(A, B)$:

$$(18) \qquad \mathbf{C}(A, B) \subset \mathbf{Rel}(A, B) = \mathfrak{P}(A \times B).$$

Inclusion (18) is the category-theoretic formulation of the Natural Law Axiom that 'Every process is a set-valued mapping.' A *model of N in the category* **C** is a member of $\mathbf{C}(N)$ with **C**-objects and **C**-morphisms.

The collections of **C**-objects and of **C**-morphisms are thus respectively of sets and of set-valued mappings, $\mathfrak{O}\mathbf{C} \subset \mathfrak{O}\mathbf{Rel}$ and $\mathfrak{A}\mathbf{C} \subset \mathfrak{A}\mathbf{Rel}$. A model of N in the category **C** may alternatively be described as a formal system that is a network of mappings in $\mathfrak{A}\mathbf{C}$, in which case one may alternately refer to 'a system N in the category **C**' when its collection $\kappa(N)$ of efficient causes is a subset of **C**-morphisms:

$$\kappa(N) \subset \mathfrak{A}\mathbf{C}. \tag{19}$$

6.12 Imminence (*RL*: 7.17) The *imminence mapping of the category* **C** (also the *imminence mapping on* **C**) is the set-valued mapping

$$\mathrm{Imm}_{\mathbf{C}} : \mathfrak{A}\mathbf{C} \mathrel{-\!\!\sqsubset} \mathfrak{A}\mathbf{C} \tag{20}$$

defined, for (set-valued mapping) $f \in \mathfrak{A}\mathbf{C}$, by

$$\mathrm{Imm}_{\mathbf{C}}(f) = \mathfrak{A}\mathbf{C} \cap \mathrm{ran}(f). \tag{21}$$

A nonempty set $\mathrm{Imm}_{\mathbf{C}}(f)$, being the collection of all **C**-morphisms that lie in the range of f, is the collection of all the f-entailed entities that can themselves entail, i.e., all possible further actions arising from f, whence the *imminence* of f. This is a key concept, indeed *the* key concept, in *RL*.

Instead of the whole collection $\mathfrak{A}\mathbf{C}$ of **C**-morphisms, consider a system N that is a network of mappings in $\mathfrak{A}\mathbf{C}$ (e.g., an (M,R)-network), whence the collection $\kappa(N)$ of all efficient causes in N is a subset of $\mathfrak{A}\mathbf{C}$, viz. $\kappa(N) \subset \mathfrak{A}\mathbf{C}$. The *imminence mapping of the system N in the category* **C** (also the *imminence mapping on* $\kappa(N)$) is the set-valued mapping

$$\mathrm{Imm}_N : \kappa(N) \mathrel{-\!\!\sqsubset} \kappa(N) \tag{22}$$

defined, for $f \in \kappa(N)$, by

$$\mathrm{Imm}_N(f) = \kappa(N) \cap \mathrm{ran}(f). \tag{23}$$

The set $\mathrm{Imm}_N(f)$ is the collection of all efficient causes of N that lie in the range of f, i.e., all the f-entailed entities in $\kappa(N)$. The imminence mapping Imm_N on $\kappa(N)$ *is* the *functional entailment pattern* of the model of the natural system N. (When $\kappa(N) = \mathfrak{A}\mathbf{C}$, e.g. when N is the whole category **C**, one has $\mathrm{Imm}_N = \mathrm{Imm}_{\mathbf{C}}$.)

6.13 Arrow Diagrams of Imminence Let $f \in \kappa(N)$ and $E = \mathrm{Imm}_N(f) \subset \kappa(N)$. The fact that $\mathrm{Imm}_N : \kappa(N) \mathrel{-\!\!\sqsubset} \kappa(N)$ is a set-valued mapping (Definition 3.1) entails the arrow diagrams

(24) $$\mathrm{Imm}_N : f \mapsto E$$

(25) $$\mathrm{Imm}_N \vdash E$$

and

(26) 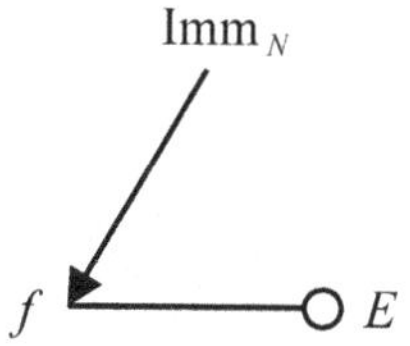

Let $g \in E = \mathrm{Imm}_N(f)$. It is evident from definition (23) of Imm_N (and that of range in Definition 1.21) that

(27) $$\begin{aligned} g \in \mathrm{Imm}_N(f) &\Rightarrow g \in \mathrm{ran}(f) \\ &\Leftrightarrow \exists x \in \mathrm{dom}(f) : g \in f(x) \subset \mathrm{ran}(f). \end{aligned}$$

That $f \in \kappa(N) \subset \mathfrak{A}\mathbf{C} \subset \mathfrak{A}\mathbf{Rel}$ is itself a set-valued mapping entails in turn the arrow diagrams

(28) $$f : x \mapsto f(x),$$

(29) $$f \vdash f(x),$$

and, since $g \in f(x)$, hereditarily (cf. Section 3.1) one has the functional entailment (*RL*: 6.14)

(30) $$f \vdash g.$$

This is to say, for $f, g \in \kappa(N)$, $g \in \mathrm{Imm}_N(f)$ if and only if f functionally entails g.

Set-valued hierarchical entailment (29) has the relational diagram

(31)

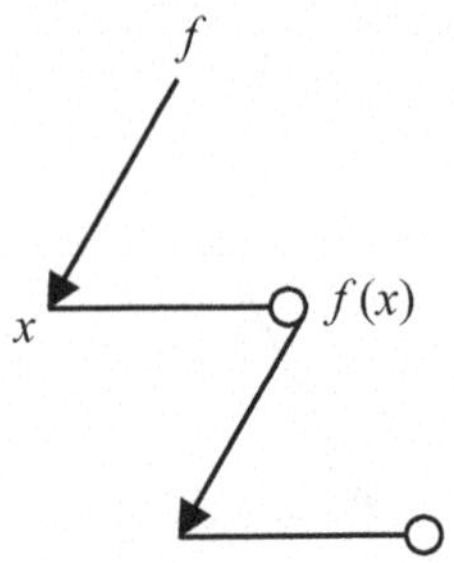

Contrast this with relational diagram (8) of the sequential composite $(g \circ f)(x)$, in which the relay vertex $f(x)$ is where the *heads* of a hollow-circle-headed arrow and a solid-headed arrow meet. Here in (31), it is the head of a hollow-circle-headed arrow and the *tail* of a solid-headed arrow that meet at the relay vertex $f(x)$. The hollow-circle heads herein just serve to denote the set-valuedness of the mappings involved, and not to distinguish sequential composition from the square product, as in relational diagrams (8) versus (10). Since $f(x)$ is a set, this connection iconography is in fact an 'abbreviation' that represents an ensemble of potentially divergent processes. All that is required is that there is *at least one* $g \in f(x)$, for which the *hierarchical composition* $f \vdash g \vdash$ proceeds, whence diagram (31) implies the canonical relational diagram of a hierarchical composition:

(32)

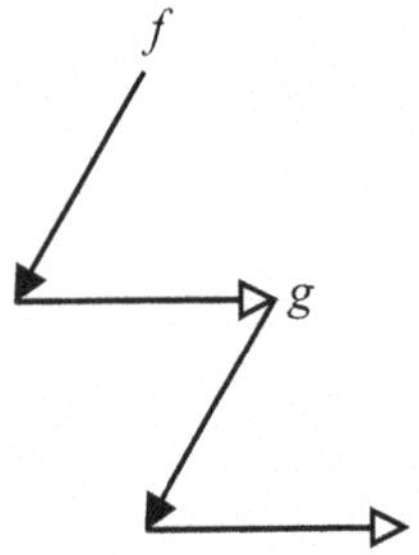

As to other members of the set $f(x)$, they may be functionally entailed, with each having its own associated $\langle$solid-headed arrow + hollow-headed arrow$\rangle$

pair, or they may be simple material outputs. For example, for a particular $f(x) = \{b_1, b_2, g_1, g_2, g_3\}$, the local element-trace network may look like this:

(33)

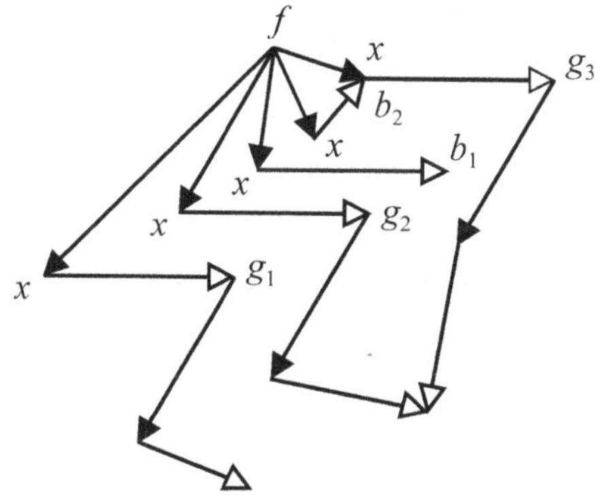

It is important to contrast mapping $f : x \mapsto f(x)$ in (28) with mapping $\mathrm{Imm}_N : f \mapsto E$ in (24). They are alternate descriptions ultimately of the functional entailment $f \vdash g$ in (30), but the two senses of 'membership' $g \in f(x)$ and $g \in E = \mathrm{Imm}_N(f)$ are hierarchically different. The mapping $f : x \mapsto f(x)$ is f *itself*, (tautologically) with domain = dom(f) and f as its *efficient* cause, while the imminence mapping $\mathrm{Imm}_N : f \mapsto E$ is *about* the entailment among processes in $\kappa(N)$, with $\mathrm{dom}(\mathrm{Imm}_N) = \kappa(N)$ and f as its *material cause.* Also, $f \vdash f(x)$ is set-valued functional entailment (in the summary interpretation of relational diagram (31)). On the other hand, $\mathrm{Imm}_N \vdash E$ may be considered material entailment of the output set E (which just so happens to be a collection of efficient causes). Indeed, in the next section we shall see how imminence Imm_N is iterated in sequential composites (while entailing iterated hierarchical composites of constituent processes).

6.14 Repair $\vdash$ Metabolism Rosen arranged the relational organization in his (M,R)-systems in such a way that *repair* is a mapping that produces as output metabolism that is itself also a mapping: for the 'R' part, instead of just producing an entity on which to operate, it could produce an operator, the 'M' part. The essence of an (M,R)-network N is the 'repair $\vdash$ metabolism' functional entailment, which may now alternatively be phrased in terms of the imminence mapping as

(34) $$\mathrm{Imm}_N : \text{repair} \mapsto \text{metabolism}.$$

In relational biology, the "What is life?" answer is given by the

Fundamental Theorem of Relational Biology (*ML*: 11.29; *RL*: 7.4)
A natural system is an organism if and only if it is closed to efficient causation.

The power of the imminence mapping formulation is immediately apparent when one sees that the very characterization of life, closure to efficient causation (*ML*: 6.23 and *RL*: 7.1–7.3), may be succinctly characterized in terms of the imminence mapping in the

6.15 Theorem (*RL*: 9.2) *A system N is closed to efficient causation if and only if, for every* $f \in \kappa(N)$, $\mathrm{Imm}_N^{-1}(f) \neq \varnothing$.

Iterated Imminence

6.16 Hierarchical Chain For $f \in \kappa(N)$, the sequential composite

(35) $$\mathrm{Imm}_N^2 = \mathrm{Imm}_N \circ \mathrm{Imm}_N : \kappa(N) \sqsubset\!\!\!- \kappa(N)$$

is defined by

(36) $$\mathrm{Imm}_N^2(f) = (\mathrm{Imm}_N \circ \mathrm{Imm}_N)(f) = \bigcup_{g \in \mathrm{Imm}_N(f)} \mathrm{Imm}_N(g) \subset \kappa(N).$$

This has the relational diagram

(37)

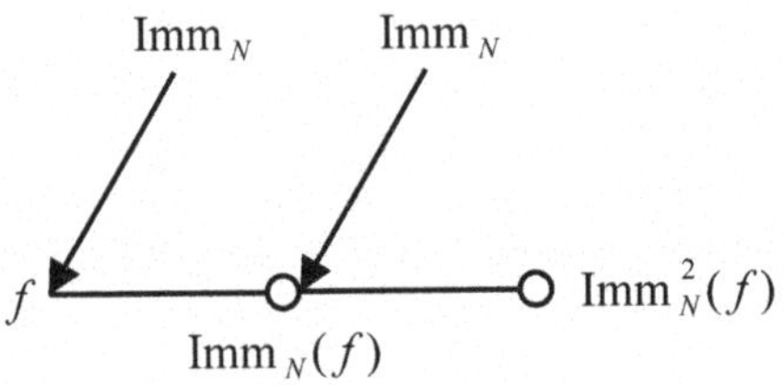

(cf. diagram (8) above). The iterated imminence $h \in \mathrm{Imm}_N^2(f)$ entails the existence of an intermediary mapping $g \in \mathrm{Imm}_N(f)$ in the imminence of f such that

(38) $$f \vdash g \vdash h.$$

Symbolically, this situation may be summarized

(39) $$h \in \mathrm{Imm}_N^2\,(f) \quad \vdash \quad (\exists g \in \mathrm{Imm}_N\,(f) : \quad f \vdash g \vdash h\ ).$$

The three mappings *f*, *g*, *h* form a hierarchical chain

(40)

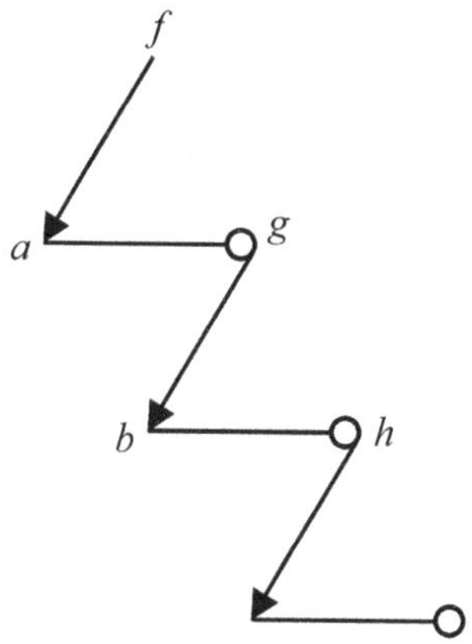

It is crucial to distinguish between the *sequential composite* $\mathrm{Imm}_N \circ \mathrm{Imm}_N$ of the set-valued imminence mapping Imm_N (relational diagram (37)) and, entailed by the iterated imminence $h \in \mathrm{Imm}_N^2(f) = \mathrm{Imm}_N \circ \mathrm{Imm}_N(f)$, the *hierarchical chain* $f \vdash g \vdash h$ among the mappings *f*, *g*, and *h* (relational diagram (40)). In category-theoretic terms, the correspondence $h \in \mathrm{Imm}_N \circ \mathrm{Imm}_N(f) \mapsto \langle f \vdash \bullet \vdash h \rangle$ is a functor that translates sequential composition into hierarchical composition. In concert with what was discussed in Section 6.13 above, here diagrams (37) and (40) are alternate descriptions of the functional entailment chain (38). But in diagram (40) *f* is an efficient cause, while 'meta-diagram' (37) is *about* the (iterated) imminence of *f*, with *f* as material cause.

6.17 Divergence Note that only the existence of one intermediary g in (38) is required, because of the '$\exists$ at least one path' characterization of the sequential composite $\mathrm{Imm}_N \circ \mathrm{Imm}_N$. The mapping f may functionally entail many more mappings in $\kappa(N)$, but none of these other branches are obliged to immediately connect to h. For example, if $g' \in \mathrm{Imm}_N(f)$ and $g' \neq g$, it may well happen that one has $h' \in \mathrm{Imm}_N(g') \subset \mathrm{Imm}_N^2(f)$, $h' \neq h$, but $h \notin \mathrm{Imm}_N(g')$. Stated otherwise, iterated imminence is *divergent*.

The iterated imminence $\mathrm{Imm}_N^2(f)$ may be interpreted as all the processes in the natural system N that are reachable from the process f after two functional entailment steps. This 'indirect imminence' is inherent in the imminence network Imm_N, and has proven to be a relational characterization of viruses (cf. *RL*: Chapter 13 on relational virology).

6.18 Convergence The square product

(41) $$\mathrm{Imm}_N \,\square\, \mathrm{Imm}_N : \kappa(N) \multimap \kappa(N)$$

is defined by

(42) $$(\mathrm{Imm}_N \,\square\, \mathrm{Imm}_N)(f) = \bigcap_{g \in \mathrm{Imm}_N(f)} \mathrm{Imm}_N(g) \subset \kappa(N).$$

Its relational diagram is

(43)

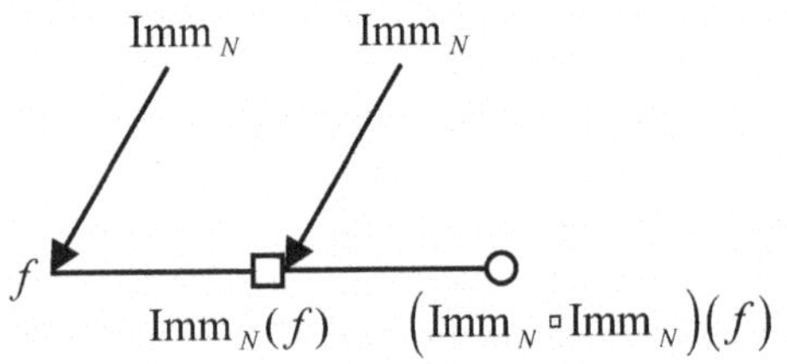

(cf. diagram (10) above). A mapping $h \in (\mathrm{Imm}_N \,\square\, \mathrm{Imm}_N)(f)$, different from the iterated imminence $h \in \mathrm{Imm}_N^2(f)$, must be reachable from the mapping f after travelling on *every* two connected arrow-pairs initiating from f in the digraph Imm_N — *all* the mappings $g_1, g_2, \ldots, g_m \in \mathrm{Imm}_N(f)$ must entail h:

(44) $$f \vdash g_1 \vdash h, \quad f \vdash g_2 \vdash h, \quad \ldots, \quad f \vdash g_m \vdash h.$$

This means a relational diagram that contains the branching pattern

(45)

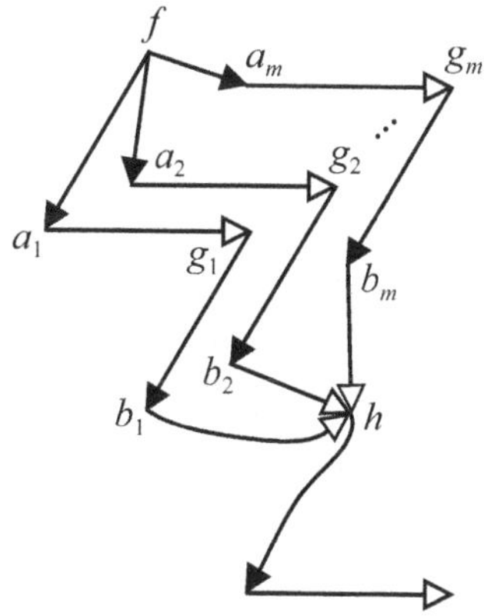

In contrast to immanent causation $\exists g$ in (39), square product imminence may symbolically be summarized as

(46) $$h \in (\mathrm{Imm}_N \,\square\, \mathrm{Imm}_N)\,(f) \quad \vdash \quad (\forall g \in \mathrm{Imm}_N\,(f) : \quad f \vdash g \vdash h\,).$$

Stated otherwise, square product imminence is *convergent*. The imposition on all intermediaries $g \in \mathrm{Imm}_N(f)$ entailed by $h \in (\mathrm{Imm}_N \,\square\, \mathrm{Imm}_N)(f)$ is an inherent *redundancy*, a multiplicity of entailment paths that says something about the importance of h to require such protection and robustness to ensure its imminent repair. Repair, in its most general terms, is set-valued functional entailment.

6.19 Moderatio The iterated imminence $h \in \mathrm{Imm}_N^2(f)$ allows a wider branching in the functional entailment network. The existence of one hierarchical composite $f \vdash g \vdash h$, that one requisite output g of f entails h, places no entailment restrictions on any other $g' \in \mathrm{Imm}_N(f)$ (so 'side-effects' may run amok in extent and depth). Contrariwise, in the square product imminence $h \in (\mathrm{Imm}_N \,\square\, \mathrm{Imm}_N)(f)$, all outputs of f (i.e. *all* effects, 'side-effects' included) are more 'reined in', mandated to at least include h in their imminence. Among the plethora of possible network connections, the generic is the 'interior' rather than the 'boundaries'. (This is the 'Goldilocks principle'.) Thus, for iterated imminence, both extremes of single-path relays ($f \vdash g \vdash h$) and all-path relays ($h \in (\mathrm{Imm}_N \,\square\, \mathrm{Imm}_N)(f)$) are exceptions, and the moderation ($h \in \mathrm{Imm}_N^2(f) = (\mathrm{Imm}_N \circ \mathrm{Imm}_N)(f)$) is the rule.

7
Metabolism and Repair

> However, it doesn't seem fruitful to reach a decision concerning the applicability of scientific concepts on sociological grounds. In general, if an individual scientist finds such concepts uncongenial, let him not use them. There is no reason why he should take their existence as a personal affront.
>
> — Robert Rosen (1977)
> in a letter to the editor of *Science*
> New Series **196**, No. 4296, *p.* 1272

Obiter dicta

7.1 Fit In the letter cited above, Robert Rosen, a stalwart in relational biology and my mentor, was commenting on the polarization, in the early days of catastrophe theory, over the immodest and hyperbolic claims made at both ends (viz. "catastrophe theory can do everything" versus "catastrophe theory can do nothing"). The quote and its lessons imparted, however, may equally apply to another esoteric subject, namely our relational biology.

Most of the antagonism against relational biology is, indeed, sociological in character. The impetuous reaction to things that one does not understand (or that one lacks the ability or knowledge to comprehend) is anger. Often, a scholar's original zeal to know was not strong enough to resist the corruption of doing what is easy and 'practical' (such as counting telephone poles). So one descends, compromises, and settles into doing routine things well. They will not be missed.

Biology is commonly perceived to be the least mathematical of the sciences, and has therefore become a safe haven for someone who would like to grow up to be a scientist but lacks the mathematical aptitude, or at least the mathematical motivation, for the 'hard sciences'. Their experimental work becomes the bulk of the observational and procedural science that is 'biology'. All mathematical matters,

A. H. Louie, *Intangible Life*, Anticipation Science 2,
DOI 10.1007/978-3-319-65409-6_8

except reluctantly 'biostatistics', are to be avoided. They may therefore convince themselves that the existence of mathematical biology, especially our intangible variety that is relational biology, is a personal affront, and as such they must conjure up petty excuses to spoil. People outside of biology, however, tend to look on us less as a threat and more as a resource for which they have been searching. They see in relational biology a language in which more enlightened alternatives to current orthodoxies may be expressed. I refer the reader to the delightful essay, a wonderful account on "hard" science and "soft" science, that Rosen reconstituted as the Praeludium of his iconoclastic masterwork *Life Itself* [Rosen 1991].

Even for the more mathematically inclined, including some 'mathematical biologists', they may reach their limits at differential equations. What practitioner needs abstract nonsense like category theory? One might be herded to the trendy social-network slogan 'Y is the new X' and formulaically proclaim 'mathematics is the new microscope', or 'biology is the next physics', and one would be exactly wrong. If one has the illusion that mathematical methods are only better mechanical instruments, then all one is equipped with are mere numerical and algorithmic tools. Likewise, one is forever hamstrung within a meagre subset of biological issues that one can possibly tackle, if one considers biology only in molecular, biochemical terms, and its problems nothing but extensions of materialistic, physical problems.

"Neque mittunt vinum novum in utres veteres; alioquin rumpuntur utres, et vinum effunditur et utres pereunt. Sed vinum novum in utres novos mittunt: et ambo conservantur." Biology and physics are different in kind. The barrier between them is that of the complex and the simple, and, as such, is not permeable. The mathematical toolbox of physics is not necessarily appropriate for biology.

7.2 Perspectives As I have explained in the Praefatio, physiochemical descriptions of biological systems in the molecular-biologic era have mostly been *internal* descriptions, which, in the Newtonian 'metric' mould, seek to express a set of states as functions of time, initial states, and the forces acting on the system. Further, the studied entities have generally been restricted to 'substances', i.e. tangible, material causes that can be isolated by conventional fractionating techniques and reductionistically studied in isolation.

Life is, however, intangible. Many biological entities, woe to reductionism, are not tangible substances, and therefore cannot be described by internal descriptions. Nicolas Rashevsky formulated the principles of relational biology to explicitly deal with them. Relational descriptions are *external* descriptions, describing the manner in which the system of interest interacts with other systems and with its environment. For example, the functional term 'active site' is a relational term, and admits only an external, relational description. External descriptions have the property that they can be realized by many distinct systems with different internal descriptions. Thus, in the case of an active site, one may imagine the same site embedded in very different kinds of physical structures (e.g., protein molecules).

In other words, relational descriptions can, characteristically, apply to a large class of functionally identical but physically quite distinct systems. Systems that exhibit the same functional properties (i.e. have the same external descriptions) are called *analogs* of one another; analogous systems are thence *alternate realizations* of the same functional system.

The very genericity of relational descriptions leads to difficulties in making connections between relational arguments and the specific metric information which constitutes the bulk of our biological knowledge. The latter refers to individual systems, while the former refer to whole classes of analogous but physically dissimilar systems. It is the role of the power set functor to connect the join of parts with the unified whole. Further, supplementing relational descriptions with appropriate optimality constraints can also overcome this general-versus-particular difficulty, and it turns out that such an optimization procedure is also functorial in character. The construction of this functor with 'optimized connections' is the subject of Part III of this monograph *IL*.

Robert Rosen devised a class of relational models called (M,R)-systems, which he introduced to the world in 1958, in his first published scientific paper [Rosen 1958a]. Indeed, (M,R)-systems were the subject of Rosen's 1959 PhD thesis (supervised by Nicolas Rashevsky, in the Committee on Mathematical Biology at the University of Chicago). The M and R may very well stand for 'metaphorical' and 'relational' in modelling terms, but they are realized as 'metabolism' and 'repair'. The comprehensive reference is [Rosen 1972] (see also *ML*: Chapters 11–13 and *RL*: Chapter 7).

Recall (Section 2.2) that the entailment of a material cause is material entailment, and the entailment of an efficient cause is functional entailment.

7.3 Metabolism Relational biology has a functional view of life, expressed in terms of processes that organisms manifest, independent of the physical substrata on which they are carried out.

An organism, being a system open to material causation (i.e., it being an open thermodynamic system), must have processes that are modes of interaction with the world. These interactions include typical material inputs that supply energy and material outputs that remove waste, and are connected together with internal processes in an entailment network that provides the capacity for renewing the structure of the organism, whatever it may be. So it is a *sine qua non* that one has to have a *metabolic* apparatus.

The word 'metabolism' comes from the Greek $\mu\varepsilon\tau\alpha\beta o\lambda\acute{\eta}$, 'change', or $\mu\varepsilon\tau\alpha\beta o\lambda\iota\sigma\mu\acute{o}\varsigma$, 'out-throw'; i.e., an alteration or a relay of materials. Metabolism, in its most general form, is thus a mapping $f : x \mapsto y$ in which $\vdash y$ is material entailment.

7.4 Repair An organism must also have a *genetic* apparatus, information carriers that tell how the products of metabolism are to be assembled. The genetic apparatus serves two functions: to *produce* the metabolic apparatus of the organism and to *reproduce* it. Rosen called the genetic processes *repair*, which, in its most general form, is a mapping $f : x \mapsto y$ in which $\vdash y$ is functional entailment.

The English word 'repair' comes from the Latin *re* + *parare*, 'make ready again'. It is, of course, a word in common usage, and means 'restore to good condition or proper functioning after damage or loss'; 'renovate or mend by replacing or fixing parts or by compensating for loss or exhaustion'; 'set right or make amends for loss, wrong, or error'. Rosen defined the technical usage of the term 'repair' in relational biology, precedently back in the beginnings of (M,R)-systems in the 1950s, to mean a hierarchical process for which 'the output of a mapping is itself a mapping'. This is the general telos of 'repair', that of an action taken to generate another action. The entailed process may possibly be previously existing, but repair does not have to be a 'return to normalcy' or 'restore to original condition'; the goal of 'the fix works' is more important.

It is unfortunate (but ultimately irrelevant) that the technical term now, alas, suffers semantic equivocation because of its usage in molecular biology to insularly mean biochemical repair of a specific molecule, that of 'DNA (and sometimes RNA) repair'. This restricted usage is a very example of the meagre appropriating the generic (analogous, for example, to the euphemism of 'period' instead of 'menstrual period', the vagueness "my system is upset" in place of "my digestive system is upset", the commercialization that the only "responsibility" is "financial responsibility", and the presumptuousness that "biology" must be "physicochemical, material-based biology"). Since the word 'repair' is not a specially coined word, its biological definition is not entitled to a universal decree. And in the absence of a default, Humpty Dumpty's rule applies: "When I use a word, it means just what I choose it to mean—neither more nor less."

7.5 Definitions To recap, our Unabashed Dictionary of Relational Biology defines

$$\text{(1)} \qquad \begin{aligned} \textit{metabolism} &= \text{material entailment,} \\ \textit{repair} &= \text{functional entailment.} \end{aligned}$$

Anything that one would want to call 'alive' would have to have at least these two basic functions of M and R, entangled in a network of interactions.

Equipped with the lessons learned in the two preceding chapters, I may also declare that M and R are natural processes that are set-valued mappings and they multifariously entail

$$\text{(2)} \qquad \begin{aligned} \textit{metabolism} &\multimap \text{(by-)products,} \\ \textit{repair} &\multimap \text{(side-)effects.} \end{aligned}$$

In (2), I have used the notation introduced in Section 6.4, that the hollow-circle-headed arrow $\multimap$ indicates that the output (final cause) of the process is a set.

(M,R)-Networks

A formal system is an object in mathematics (*ML*: 4.6 and Chapter 7). One may, without loss of generality, simply consider a formal system as a set S with a collection $\kappa(S)$ of mappings; so a formal system is the ordered pair $\langle S, \kappa(S)\rangle$. The collection $\kappa(S)$ is a collection of interconnected arrows that represent the inferential patterns.

7.6 Definition (*ML*: 13.2; *RL*: 7.10) *Metabolism* and *repair* are input-output systems that are connected as *components* into a network. They are formal systems with the following further category-theoretic structures.

i. A *metabolism component* is a formal system $M_i = \langle A_i, H(A_i, B_i)\rangle$.
ii. A *repair component* is a formal system $R_i = \langle Y_i, H(Y_i, H(A_i, B_i))\rangle$.
iii. A *metabolism-repair network*, i.e., an *(M,R)-network*, is a finite collection of pairs of metabolism and repair components $\{(M_i, R_i) : i \in I\}$, connected in a model network. In particular, the output of a repair component R_i are observables in $H(A_i, B_i)$ of its corresponding metabolism component M_i. The metabolism components may be connected among themselves by their inputs and outputs (i.e., by $B_k \subset A_j$ for some $j, k \in I$). Repair components must receive at least one input from the outputs of the metabolism components of the network (i.e., $Y_i = \prod_{k=1}^{n} B_{i_k}$ with $n \geq 1$ and where each $i_k \in I$).

Note that the connections specified in iii are the *requisite* ones; an (M,R)-network may have additional interconnections among its components and with its environment.

7.7 Definition (*ML*: 11.13; *RL*: 7.11) An *(M,R)-system* is an (M,R)-network that is closed to efficient causation.

How an (M,R)-network achieves this closure to become an (M,R)-system is the subject of the next chapter.

(M,R)-systems began as a class of metaphorical, relational paradigms that define cells. It is, however, not much of a hyperbole to declare that all of Rosen's scientific work—his lifelong quest being the answer to the question "What is life?"—has arisen from a consideration of topics related to the study of (M,R)-systems. This is because of the

7.8 Postulate of Life (*ML*: 11.28, *RL*: 8.30) A natural system is an organism if and only if it realizes an (M,R)-system.

Here, the word 'organism' is used in the sense of a general living system (including, in particular, cells). Thus an (M,R)-system is the very model of life; and, conversely, life is the very realization of an (M,R)-system.

A union of interacting (M,R)-systems (or better, their *join* in the *lattice* of (M,R)-systems; cf. *ML*: 2.1 and 7.28) is itself an (M,R)-system. A multicellular organism has a life of its own, apart from the fact that the cells that comprise it are alive. Similarly, in some sense an ecosystem of interacting organisms is itself an organism. In particular, a symbiotic union of organisms may itself be considered an organism (*RL*: 11.12).

For the record, I reiterate the fact that an organism is *closed to efficient causation*, but *open to material causation*.

$\mathbf{M} \cap \mathbf{R} \neq \varnothing$

Metabolism, in its most general terms, is set-valued material entailment. 'M' is all the material entailments that occur within a living organism, a relay network of metabolites. The *telos* of metabolism is energy production, but with by-products galore.

C continues to be a subcategory of **Rel** from which I draw models of a natural system N.

7.9 Global Bundle The *metabolism bundle* (*of the category* **C**) is the set-valued mapping

$$\text{(3)} \qquad \mathrm{Met}_{\mathbf{C}} : \mathfrak{A}\mathbf{C} \multimap \mathfrak{A}\mathbf{C}$$

defined by

$$\text{(4)} \qquad \mathrm{Met}_{\mathbf{C}} = \{\, (f,g) \in \mathfrak{A}\mathbf{C} \times \mathfrak{A}\mathbf{C} : \mathrm{dom}(g) \cap \mathrm{ran}(f) \neq \varnothing \,\}.$$

The subset $\mathrm{Met}_{\mathbf{C}} \subset \mathfrak{A}\mathbf{C} \times \mathfrak{A}\mathbf{C}$ is the domain on which 'metabolism' in **C** may proceed, containing pairs of processes (f,g) that may participate in the relay $x \mapsto f(x) \mapsto g(f(x))$. Hence the expression 'Met' as the symbol, and the name metabolism bundle that I have given it. (For an explanation of the usage of 'bundle', see *RL*: 10.5.)

The analogy between Met and Imm (Definition 6.12) is more apparent if I rephrase the set-valued mapping $\mathrm{Imm}_{\mathbf{C}}$ also as a subset of $\mathfrak{A}\mathbf{C} \times \mathfrak{A}\mathbf{C}$:

(5)
$$\begin{aligned}\mathrm{Imm}_{\mathbf{C}} &= \{(f,g) \in \mathfrak{A}\mathbf{C} \times \mathfrak{A}\mathbf{C} : g \in \mathrm{ran}(f)\} \\ &= \{(f,g) \in \mathfrak{A}\mathbf{C} \times \mathfrak{A}\mathbf{C} : \{g\} \cap \mathrm{ran}(f) \neq \varnothing\}\end{aligned}.$$

7.10 Local Bundle Similar to the imminence mapping, the metabolism bundle may likewise be restricted to a model of a natural system N in the category **C**, as a 'model-specific' set-valued mapping

(6)
$$\mathrm{Met}_N : \kappa(N) \multimap \kappa(N)$$

defined by

(7)
$$\mathrm{Met}_N = \{(f,g) \in \kappa(N) \times \kappa(N) : \mathrm{dom}(g) \cap \mathrm{ran}(f) \neq \varnothing\},$$

which is the embodiment of the material entailment structure in N.

7.11 Dual Roles The two subsets Met_N and Imm_N of $\kappa(N) \times \kappa(N)$, i.e., *metabolism* and *repair* in the natural system N, are not necessarily disjoint. The range of a mapping may contain both materially-entailed and functionally-entailed entities. A single output set of a set-valued mapping may itself already contain both species. It may also happen that a single output entity takes on *dual roles* of being materially entailed in one interaction and functionally entailed in another.

7.12 Triumvirate Revisited As an illustrative example of the latter, consider the simplest (M,R)-system (cf. *ML*: Chapters 11 and 12). Its relational diagram is

(8)

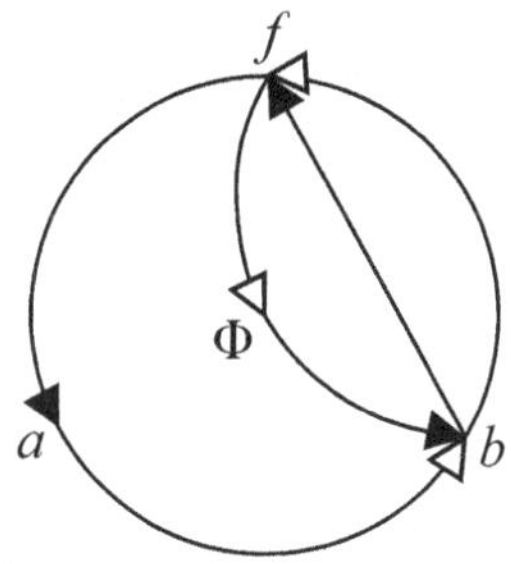

The subgraph

(9)

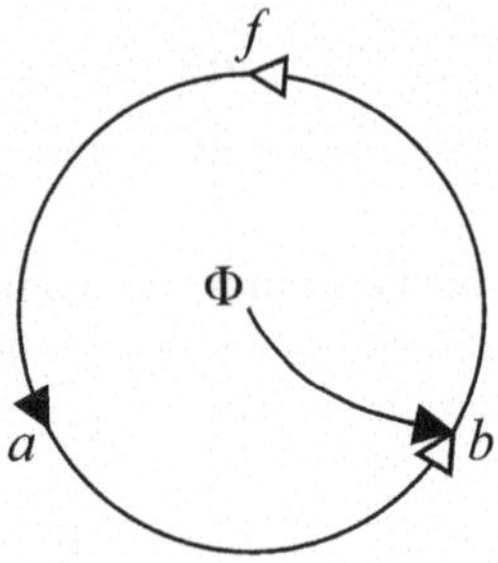

shows the hierarchical composition of the metabolism mapping $f : a \mapsto b$ and its repair mapping $\Phi : b \mapsto f$, $\Phi \vdash f$ being the very definition of repair:

(10)

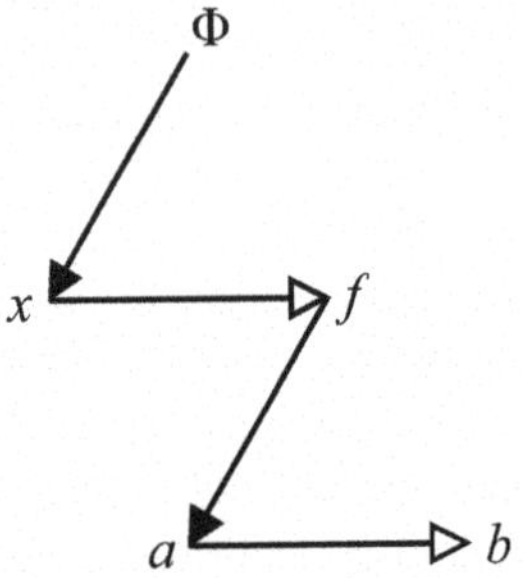

whence f is functionally entailed by Φ and relayed in

(11) $$\Phi \vdash f \vdash b.$$

Since the repair map Φ uses as input the output b of the metabolism map f, diagram (9) also shows the metabolism map $f : a \mapsto b$ and the repair map $\Phi : b \mapsto f$ in *sequential* composition:

(12)

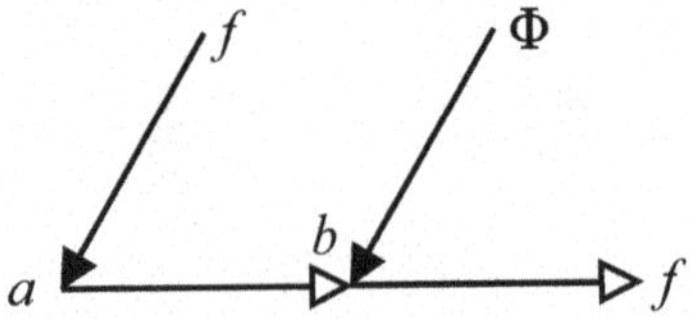

whence b is materially entailed by f and relayed in

(13) $$\Phi \circ f : a \mapsto b \mapsto f.$$

The subgraph

(14)

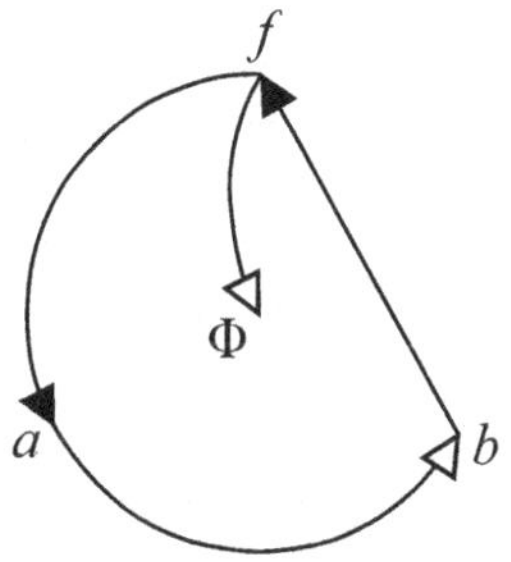

of (8) shows the metabolism map $f : a \mapsto b$ and the replication map $b : f \mapsto \Phi$ in *hierarchical* composition:

(15)

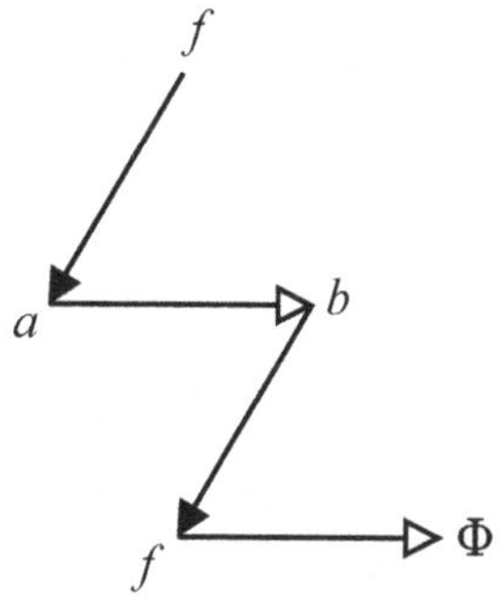

whence b is functionally entailed by f and relayed in

(16) $$f \vdash b \vdash \Phi$$

So one sees that the output entity b takes on dual roles, that of a materially entailed product in (9)–(13), and of a functionally entailed effect in (14)–(16).

In relational diagrams (8), (9), (12), (14), and (15), some hollow-triangle-headed arrows may be replaced by hollow-circle-headed arrows or a hollow-square-headed arrows (as introduced in Section 6.4) to emphasize the involved compositions.

The Nuances of Repair Action

7.13 Repair $\vdash$ Metabolism The sequential composition (9) and (12) of the metabolism map $f : a \mapsto b$ and the repair map $\Phi : b \mapsto f$ may be represented by the diagram

(17) $$A \xrightarrow{f} B \xrightarrow{\Phi} H(A,B)$$

that shows the domains and codomains of the mappings.

The mapping f represents *metabolism*, whence its efficient cause, an *enzyme*, with material input and output represented by the sets A and B. Thus metabolism is a morphism

(18) $$f \in H(A,B) \subset B^A.$$

Members of $H(A,B)$ are mappings that model metabolic processes, so clearly not all mappings in B^A qualify; $H(A,B)$ is therefore a proper subset of B^A.

The mapping Φ represents *repair*. Its codomain is $H(A,B)$, so it may be considered as a mapping that creates new copies of enzymes f, hence a *gene* that 'repairs' the metabolism function. In other words, repair is a morphism Φ with the prescribed codomain $H(A,B)$; i.e. $\Phi \in H(\bullet, H(A,B)) \subset H(A,B)^{\cdot}$. Repair in cells generally takes the form of a continual synthesis of basic units of metabolic processor (i.e. enzymes), using as inputs materials provided by the metabolic activities themselves. Stated otherwise, the domain of the repair map Φ is the codomain of metabolism f, its 'output set' B. Thus

(19) $$\Phi \in H(B, H(A,B)) \subset H(A,B)^B.$$

In the multi-component general (M,R)-system, a metabolism mapping would be $f \in H(A_i, B_i)$, and a repair mapping would be $\Phi \in H(Y_i, H(A_i, B_i))$ with $Y_i = \prod_{k=1}^{n} B_{i_k}$, as in Definition 7.6. The repair map $\Phi : b \mapsto f$, the processor that creates new copies of enzymes $f \in H(A_i, B_i)$, using as input material $b \in \prod_{k=1}^{n} B_{i_k}$ provided by the output(s) of metabolic activities, is the functional entailment that one may *dogmatically* declare as the *central* efficient cause of life, the prototypical "selfish gene" by another name.

7.14 Functional Entailment as Set-Valued Mapping Robert Rosen famously declared the biological realization of functional entailment his deepest insight. The innovation is that the processes themselves may be treated like any other output object. In his (M,R)-systems, Rosen defined repair to be a mapping that produces metabolism, an output that is itself also a mapping. An 'R' component, instead of entailing an entity on which to be processed, entails a process that is an 'M' component. In short, R $\vdash$ M.

That hierarchical composition is *formally* different from sequential composition is evident from a comparison of the entailment chains (11) and (13) (and their corresponding arrow diagrams (10) and (12)). But functional entailment and material entailment are not *categorically* different. One reason is the natural isomorphism

(20) $$H(X \times A, B) \cong H(X, H(A, B))$$

(cf. *RL*: 6.14), so a repair map $\Phi \in H(X, H(A, B))$ that entails $f \in H(A, B)$ may be equivalently represented as a material-entailing mapping from $X \times A$ to B. (The natural isomorphism (20) is of great importance in relational biology. It will reappear here in *IL* in Sections 8.9 et seq., and Sections 11.1 et seq.) Another reason is that both kinds of entailments may be considered set-valued mappings, whence elements of the hom-set $\mathbf{Rel}(X, Y) \cong \mathbf{Svm}(X, Y)$ for appropriate sets X and Y. Indeed, a metabolism map $f \in H(A, B)$ is

(21) $$f \in \mathbf{Set}(A, B) \subset \mathbf{Rel}(A, B),$$

and a repair map $\Phi \in H(X, H(A, B))$ is a fortiori

(22) $$\begin{aligned}\Phi \in \mathbf{Set}(X, B^A) &\subset \mathbf{Set}(X, \mathrm{P}(A \times B)) \\ &= \mathbf{Svm}(X, A \times B) \cong \mathbf{Rel}(X, A \times B)\end{aligned}.$$

7.15 Repair Action on Its Material Cause The relational diagram

(23)

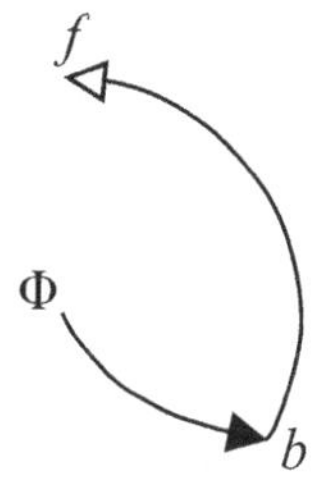

is a succinct representation of a multitude of subprocesses, which may be partitioned into a material first half and a functional second half:

$$\Phi : \{\, b \,\} \longrightarrow [\![\, b \,]\!] \longrightarrow f. \tag{24}$$

The material first half takes as its material cause $\{\, b \,\}$, a collection of amino acids, as input, and processes it into a polypeptide $[\![\, b \,]\!]$. The involved subprocesses, the $\langle$ efficient cause : material cause $\rangle$ interaction, $\langle\, \Phi : b \,\rangle$, is realized, in all its molecular-biologic glory, in the action chain:

(25)

$\Phi \vdash$ transcription
(from the 4-letter-alphabet DNA-language to
the slightly variant 4-letter-alphabet RNA-dialect)
$\vdash$ transcription
(from the 4-letter-alphabet RNA-language to
the 20-letter-alphabet protein-language,
using the $4^3 \to 20$ genetic code) : $\{b\}$
$\vdash$ polymerization into a polypeptide : $[\![b]\!]$.

7.16 Repair Entails Its Final Cause The functional second half of the entailment $\Phi \vdash f$ then continues, endowing the inert polypeptide chain $[\![\, b \,]\!]$ with its ultimate enzymatic activity as an efficient cause f, as follows:

(26)

$\Phi \vdash$ post-translational modification of $[\![\, b \,]\!]$
$\vdash$ folding into native state of the protein molecule
$\vdash$ protein as enzyme has active site(s)
$\vdash$ active site provides function
$\vdash$ enzymatic activity of the output $\mapsto f$.

Note that while the steps in the action sequence (25) are now quite well-understood thanks to the past 60 years of success of molecular biology, the steps in the action sequence (26) are not. These latter steps have to do with how structures attain functions, and the processes are all ‘complex’ (in the *impredicative* sense), and I submit that their explications require relational-biologic tools, some of which will be offered in the next chapter.

7.17 Ubiquity Final causes of processes are not ends in themselves. In the entailment network $\kappa(N)$ of a natural system N, processes are in composition, and among them final causes are further relayed as material and efficient causes. The entailment network $\kappa(N)$ is completely described by two set-valued mappings defined on it: the metabolism bundle Met_N generates (by-)products through material entailment, and the imminence mapping Imm_N generates (side-) effects through functional entailment. Every natural process in $\kappa(N)$ may be categorized as either 'metabolism' or 'repair', even when N is not necessarily a metabolism–repair network per se. Together, Met_N and Imm_N may be taken as the very definition of the entailment network that models the natural system N. As we have just seen in the (M,R)-system example, the two set-valued mappings are not necessarily mutually exclusive. After all, by-products and side-effects are but different names that denote the same multifarious outputs, those which are entailed, of interacting processes.

Metabolic Entailment Between Systems

The efficient cause of entailment can just as commonly arise without as within.

7.18 Symbiosis Redux Consider two formal systems (which may, but not necessarily, be (M,R)-networks) $\langle H, \kappa(H)\rangle$ and $\langle S, \kappa(S)\rangle$; that is, systems H and S with their respective collections $\kappa(H)$ and $\kappa(S)$ of efficient causes. Two systems interact when a process in one system affects another system. Stated otherwise, an interactive connection $S \to H$ happens when the final cause of a process in $\kappa(S)$ is further relayed in H. The theme of *RL* is "How do two lifeforms interact?". One ubiquitous biological interaction is *symbiosis* (*RL*: Chapter 11), between a *host* and a *symbiont*. This is the source of the symbols H and S. One may use host–symbiont interaction as a running example of the system interactions now under consideration.

7.19 Metabolic Relays Let two systems H and S be modelled in a category **C**, whence the two sets of efficient causes (i.e. two collections of set-valued mappings) $\kappa(H)$ and $\kappa(S)$ are subsets of $\mathfrak{A}\mathbf{C}$. Let g and f be processes in H and S respectively; i.e. $g \in \kappa(H)$ and $f \in \kappa(S)$. If $x \in \mathrm{dom}(f)$ is such that $\mathrm{dom}(g) \cap f(x) \neq \varnothing$, then $\exists\, y \in f(x)$ and the relay

$$x \mapsto y \mapsto g(y) \tag{27}$$

can proceed.

I shall use the canonical hierarchical cycle

(28) 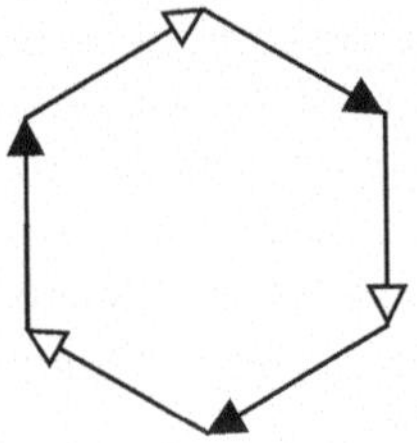

to symbolically represent the system H, and a single arrow pair

(29) 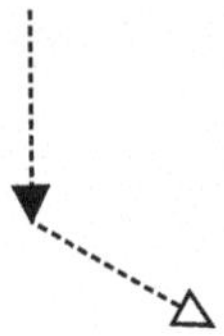

to denote the system S. Of course, the entailment networks of both H and S are far more complicated relational diagrams consisting of large numbers of interconnected arrows. But the interactions between diagrams (28) and (29) are sufficient to illustrate the modes of interactions that I shall discuss. Also, I shall be varying the hollow 'arrowheads' (triangle for element-trace, circle for sequential composition, and square for square product) to represent the different entailment patterns.

The elemental relay (27) is the H–S interaction in the *join* entailment network $N = H \vee S$ (cf. *ML*: 2.1 and 7.28 and *RL*: 11.12 and 13.2):

(30) 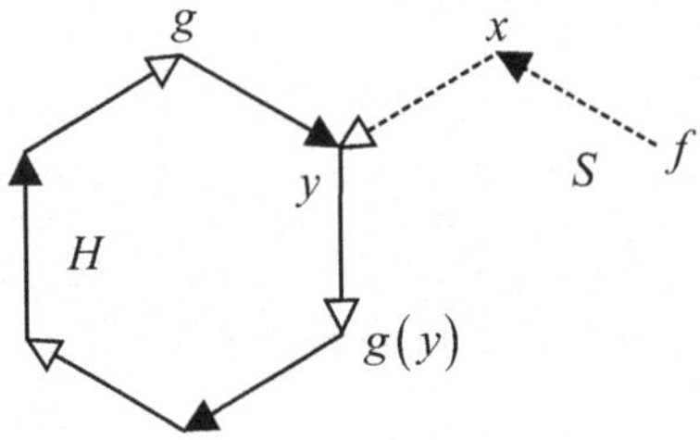

In Chapter 11 of *RL*, this mode of interaction between H and S is explicated as symbiosis (when the metabolite y is shared between the two systems) and alternatively as infection (when the antigen $y \in f(x)$ produced by S invades the host H). The elemental relay (27) is also an instance of sequential composition of the two mappings g and f. Depending on the entailment paths involved, it can either be the sequential composite $g \circ f$:

(31)

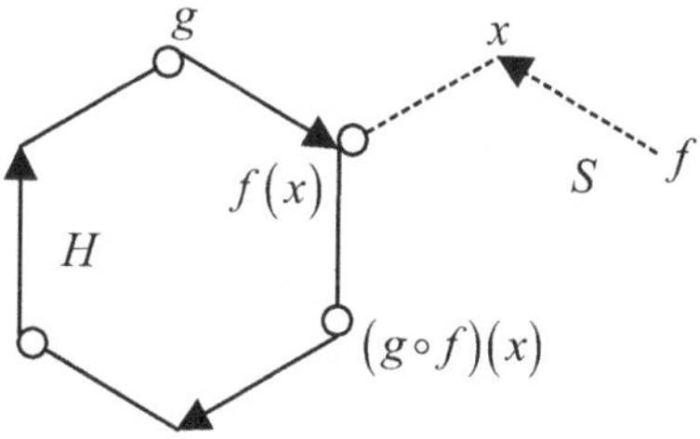

or the square product $g \square f$:

(32)

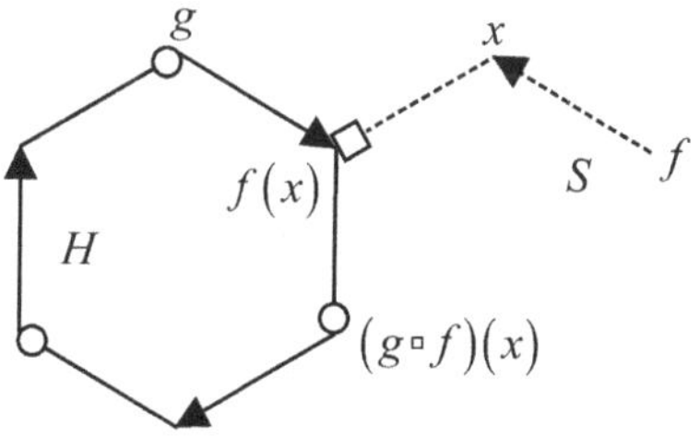

These two relational diagrams (31) and (32) then serve as models of metabolic interactions between H and S when by-products are involved. Suppose in the original system H, before its interaction with S, the entailment of the mapping $g \in \kappa(H)$ is

(33) $$g \vdash z$$

(with $z \neq g(y)$).

The metabolic entailments of the mapping $g \in \kappa(H \vee S)$ in relational diagrams (30), (31), and (32) are, respectively,

(34) $$g \vdash g(y), \quad g \vdash (g \circ f)(x), \quad \text{and} \quad g \vdash (g \square f)(x).$$

The number of entailment paths involved in the sequential composition of the two mappings g and f increases in these three entailments in (34), from one to at least one to all. This difference in degree may then be realized as a measure of the ease of implementing therapeutic procedures to revert the affected $g \in \kappa(H \vee S)$ back to its native state $g \in \kappa(H)$ with entailment (33).

7.20 Metabolic Action $\mathrm{Met}_{S \to H}$ How the two mappings $g \in \kappa(H)$ and $f \in \kappa(S)$ sequentially compose is completely determined by the join entailment network $\kappa(N) = \kappa(H \vee S)$. Instead of the metabolism bundle $\mathrm{Met}_N \subset \kappa(N) \times \kappa(N)$, one may consider its restriction to the two interacting subsystems H and S. The *metabolism bundle of the action of S on H*,

$$(35) \qquad \mathrm{Met}_{S \to H} : \kappa(S) \dashv\sqsubset \kappa(H),$$

takes the form

$$(36) \qquad \mathrm{Met}_{S \to H} = \{(f, g) \in \kappa(S) \times \kappa(H) : \mathrm{dom}(g) \cap \mathrm{ran}(f) \neq \varnothing\}.$$

The entailment network $\mathrm{Met}_{S \to H} \subset \kappa(S) \times \kappa(H)$ contains all the metabolic consequences of S on H.

7.21 Effects $\kappa(S \to H)$ Note the inclusions $\kappa(S) \times \kappa(H) \subset \kappa(N) \times \kappa(N) \subset \mathfrak{A}\mathbf{C} \times \mathfrak{A}\mathbf{C}$, whence the three metabolism bundles on different domains are related thus:

$$(37) \qquad \mathrm{Met}_{S \to H} = \mathrm{Met}_N|_{\kappa(S) \times \kappa(H)} \quad \text{and} \quad \mathrm{Met}_N = \mathrm{Met}_{\mathbf{C}}|_{\kappa(N) \times \kappa(N)}.$$

If $(f, g) \in \mathrm{Met}_{S \to H}$, then a material relay $x \mapsto f(x) \mapsto g(f(x))$ may be defined on $X_g = \{x \in \mathrm{dom}(f) : f(x) \in \mathrm{dom}(g)\}$. But this restriction $g \circ f|_{X_g}$ may not necessarily be expandable to the sequential composite $g \circ f$ on all of $\mathrm{dom}(f)$, and it may not be in the existing collections $\kappa(H)$ or $\kappa(S)$ of processes. The mapping $g \circ f|_{X_g}$ arises from the interaction. If one denotes the effects of S on H (i.e., the collection of processes in the interaction $S \to H$) by $\kappa(S \to H)$, then $g \circ f|_{X_g} \in \kappa(S \to H)$.

Functional Entailment Between Systems

7.22 Inter-Network Imminence $\mathrm{Imm}_{S\to H}$ The imminence mapping $\mathrm{Imm}_N \subset \kappa(N) \times \kappa(N)$ of the system $N = H \vee S$ may likewise be restricted to the two interacting subsystems H and S. Define the value of the set-valued mapping

(38) $$\mathrm{Imm}_{S\to H} : \kappa(S) \multimap \kappa(H)$$

at $f \in \kappa(S)$ as

(39) $$\mathrm{Imm}_{S\to H}(f) = \kappa(H) \cap \mathrm{ran}(f).$$

Hierarchical composition $f \vdash g$ may be defined for $f \in \kappa(S)$ and $g \in \kappa(H)$ if and only if

(40) $$g \in \kappa(H) \cap \mathrm{ran}(f) = \mathrm{Imm}_{S\to H}(f) \neq \varnothing.$$

The set $\mathrm{Imm}_{S\to H}(f)$ contains all the processes in the system H that may be functionally entailed by the process $f \in \kappa(S)$ of the system S. In other words, $\mathrm{Imm}_{S\to H}(f)$ contains all possible further actions in the system H arising from interacting with $f \in \kappa(S)$. The set-valued mapping $\mathrm{Imm}_{S\to H}$ may, therefore, be considered the *imminence of S on H*, i.e., *inter-network imminence*. Functional entailment is *repair* in its most general sense, whence $\mathrm{Imm}_{S\to H}(f) \subset \kappa(S \to H)$ may be considered a *repair effect* in the interaction $S \to H$.

Similar to the three metabolism bundles, the three imminence mappings on different domains are related thus:

(41) $$\mathrm{Imm}_{S\to H} = \mathrm{Imm}_N|_{\kappa(S)\times\kappa(H)} \quad \text{and} \quad \mathrm{Imm}_N = \mathrm{Imm}_{\mathbf{C}}|_{\kappa(N)\times\kappa(N)}.$$

7.23 Pathophysiology For $f \in \kappa(S)$ and $g \in \kappa(H)$, the functional entailment $f \vdash g$, i.e.

(42) $$g \in \mathrm{Imm}_{S\to H}(f),$$

has the relational diagram

(43)

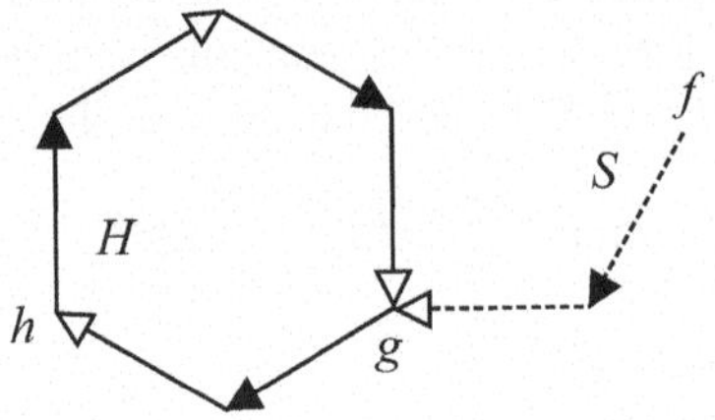

This interaction, the hierarchical composite

(44) $f \vdash g \text{ in } S\,,\quad g \vdash h \text{ in } H \quad \Rightarrow \quad f \vdash g \vdash h \text{ in } N = H \vee S,$

models competing processes between H and S, as well as repair-level and replication-level infections of H by S (*RL*: Chapters 11 and 13). The process $h \in \kappa(H) \subset \kappa(H \vee S) = \kappa(N)$ is reachable from the process $f \in \kappa(N)$ after two functional entailment steps. Thus, depending on the entailment paths involved, it can be formulated in terms of an iterated imminence as either the sequential composite

(45) $$h \in \mathrm{Imm}_N^2(f)$$

or the square product

(46) $$h \in (\mathrm{Imm}_N \,\square\, \mathrm{Imm}_N)(f)$$

(cf. the discussions on iterated imminence in the previous chapter, Sections 6.16–6.19). These two relational diagrams (45) and (46) then serve as models of functional interactions between H and S when side-effects are involved. The original functional entailment

(47) $$g_0 \vdash h_0$$

in the system H

(48)

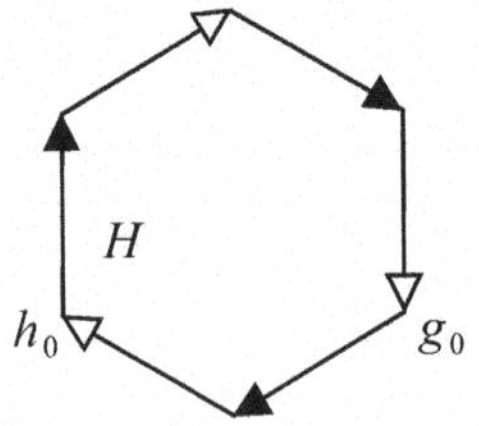

is perturbed by the interaction with S. With increasing number of entailment paths in the hierarchical composition of $f \vdash g \vdash$, the functional entailments involved in (44), (45), and (46) are, respectively,

$$(49) \qquad f \vdash g \vdash h,$$

$$(50) \qquad f \vdash \mathrm{Imm}_N(f) \vdash h \in \mathrm{Imm}_N^2(f),$$

and

$$(51) \qquad f \vdash \mathrm{Imm}_N(f) \vdash h \in (\mathrm{Imm}_N \circ \mathrm{Imm}_N)(f).$$

Synthesis

7.24 Join When two formal systems $\langle H, \kappa(H)\rangle$ and $\langle S, \kappa(S)\rangle$ interact, their entailment networks connect to become the *join* formal system $\langle H \vee S, \kappa(H \vee S)\rangle$ (*RL*: 13.2). The material base set of $H \vee S$ is quite straight-forwardly $H \cup S$, but the collection $\kappa(H \vee S)$ of join processes is more than the union $\kappa(H) \cup \kappa(S)$. This is because, in addition to the processes $\kappa(H)$ and $\kappa(S)$ within the two systems, join processes in $\kappa(H \vee S)$ must also include the mutual interactions between H and S: the effects $\kappa(S \to H)$ of S on H, and the effects $\kappa(H \to S)$ of H on S. Thus

$$(52) \qquad \kappa(H \vee S) = \kappa(H) \cup \kappa(S) \cup \kappa(S \to H) \cup \kappa(H \to S).$$

7.25 Corange Unions Interactive processes between H and S may be *synthesized* from the set-valued mappings Met and Imm. Note that

$$(53) \qquad \begin{array}{ll} \mathrm{cor}(\mathrm{Met}_{S\to H}) \subset \kappa(S), & \mathrm{cor}(\mathrm{Imm}_{S\to H}) \subset \kappa(S); \\ \mathrm{cor}(\mathrm{Met}_{H\to S}) \subset \kappa(H), & \mathrm{cor}(\mathrm{Imm}_{H\to S}) \subset \kappa(H). \end{array}$$

The corange

$$(54) \qquad \mathrm{cor}(\mathrm{Met}_{S\to H}) = \{f \in \kappa(S) : \exists g \in \kappa(H)\ \ \mathrm{dom}(g) \cap \mathrm{ran}(f) \neq \varnothing\}$$

contains all the processes in $\kappa(S)$ that produce metabolism effects in H. Likewise, $\mathrm{cor}(\mathrm{Imm}_{S\to H})$ contains all the processes in $\kappa(S)$ that produce repair effects in H. Every process may function as either 'metabolism' or 'repair' (or a combination thereof), so the union of material entailment and functional entailment $\mathrm{cor}(\mathrm{Met}_{S\to H}) \cup \mathrm{cor}(\mathrm{Imm}_{S\to H})$ completely describes the effect of $\kappa(S)$ on $\kappa(H)$. Let me introduce the notation

(55) $$[\kappa(S) -\!\!\sqsubset \kappa(H)] = \mathrm{cor}(\mathrm{Met}_{S\to H}) \cup \mathrm{cor}(\mathrm{Imm}_{S\to H}).$$

Conversely, $\mathrm{cor}(\mathrm{Met}_{H\to S})$ and $\mathrm{cor}(\mathrm{Imm}_{H\to S})$ are the metabolism and repair effects of $\kappa(H)$ on S, whence

(56) $$[\kappa(H) -\!\!\sqsubset \kappa(S)] = \mathrm{cor}(\mathrm{Met}_{H\to S}) \cup \mathrm{cor}(\mathrm{Imm}_{H\to S}).$$

7.26 Approximation The best approximation of the collection of join processes in $H \vee S$ is then the union of the active processes in $\kappa(H)$ and $\kappa(S)$ with these four coranges:

(57) $$\begin{aligned}\kappa(H \vee S) \approx \kappa(H) &\cup \kappa(S)\\ &\cup \mathrm{cor}(\mathrm{Met}_{S\to H}) \cup \mathrm{cor}(\mathrm{Imm}_{S\to H})\\ &\cup \mathrm{cor}(\mathrm{Met}_{H\to S}) \cup \mathrm{cor}(\mathrm{Imm}_{H\to S})\ ;\end{aligned}$$

that is,

(58) $$\begin{aligned}\kappa(H \vee S) \approx \kappa(H) &\cup \kappa(S)\\ &\cup [\kappa(S) -\!\!\sqsubset \kappa(H)] \cup [\kappa(H) -\!\!\sqsubset \kappa(S)].\end{aligned}$$

The set-valued mappings Met and Imm are mappings of *potentiality*. They trace the possible material and functional entailments arising from a system, i.e., the system's possible metabolism and repair effects. This propensity for the emergence of material and functional entailments inherent in Met and Imm is what allows the synthetic continuation from $\kappa(H)$ and $\kappa(S)$ to $\kappa(S \to H)$ and $\kappa(H \to S)$. Note, however, that one can only reconstruct the interactive processes between H and S from processes that already exist (but are dormant) in the partitioned $\kappa(H)$ and $\kappa(S)$. Note the containments (53). A process in $\kappa(H \vee S)$ that becomes extinct in the fractionation of $\kappa(H \vee S)$ into $\kappa(H)$ and $\kappa(S)$ cannot be recovered through Met and Imm. Stated otherwise,

$$(59) \qquad [\kappa(S) \;\neg\!\sqsubset\; \kappa(H)] = \mathrm{cor}(\mathrm{Met}_{S\to H}) \cup \mathrm{cor}(\mathrm{Imm}_{S\to H}) \subset \kappa(S \to H)$$

and

$$(60) \qquad [\kappa(H) \;\neg\!\sqsubset\; \kappa(S)] = \mathrm{cor}(\mathrm{Met}_{H\to S}) \cup \mathrm{cor}(\mathrm{Imm}_{H\to S}) \subset \kappa(H \to S),$$

and both containments may be proper. Thus the unions (57) and (58) are only an *approximation* of the union (52), but this is the best effort in the synthesis of the latter sum from the analytic parts $\kappa(H)$ and $\kappa(S)$.
This is yet another example of

$$(61) \qquad \text{synthesis} \circ \text{analysis} \not\equiv \text{identity},$$

the non-invertibility I have already discussed in Sections 1.22 (cf. also *ML*: 7.43–7.49). Fractionation is almost always irreversible for complex systems; that is why the reductionistic strategy of understanding by parts is inherently flawed. The gathering of all possible information in order to reconstruct from calamity, to best undo disastrous fractionation, is an art.

Therapeutics

7.27 Material Therapeutics The plurality of process outputs is a genericity of Nature. All actions have consequences, most of them being side-effects. Therapeutics (medical or otherwise) is the art of undoing actions.

For an illustrative example, consider an attempt in the restoration to the native state (48) of the system H from the interactive state (43) of the join system $N = H \vee S$. A treatment scheme that involves nothing else but the reversal from the perturbed $g \vdash h$ in (44) back to the native $g_0 \vdash h_0$ in (47) is relatively straightforward. But, unfortunately, the singular elemental-tracing relay $f \vdash g \vdash h$ in (44) is a simplified entailment model of the generic hierarchical compositions of set-valued processes that are either (45) or (46). Reductionistic biochemical therapeutics operates on the level of

$$(62) \qquad f \vdash g \vdash h \quad \to \quad g_0 \vdash h_0.$$

This path-tracing model, the specification of one intermediary process $g \in \mathrm{Imm}_N(f)$, misses all the other processes in $\mathrm{Imm}_N(f)$. With only singular $g \in \mathrm{Imm}_N(f)$ and $h \in \mathrm{Imm}_N^2(f)$ (or possibly $h \in (\mathrm{Imm}_N \,\square\, \mathrm{Imm}_N)(f)$) on hand, too much information is lost for the recovery of $g_0 \vdash h_0$. Often, one does not have information on the directly-affected g, nor, indeed, on the iterated imminence h. The actual 'observables' ('symptoms') may be some indirect effects several iterations of hierarchical compositions further along the causal chain.

7.28 Relational Therapeutics Relational therapeutics suggests

$$(63) \qquad h \in \mathrm{Imm}_{S \to H}^2(f) \quad \to \quad h_0 \in \mathrm{Imm}_H(g_0),$$

on a different hierarchical level.

The relational therapy (63) hinges on knowledge of both the imminence mappings of the healthy state Imm_H and the pathophysiology $\mathrm{Imm}_{S \to H}$, or equivalently, the entailment networks $\kappa(H)$ and $\kappa(H \vee S)$, in which every process is a set-valued mapping. Success depends on the size of the available fragment of imminence. Note the disentanglement suggested in (63) is not the panacea of an 'inverse operation' to 'neutralize' the effect of $f \in \kappa(S)$ on H. Once imminence has been released, laments Lady Macbeth, "What's done cannot be undone." Even in the unlikely scenario of complete information (that is to say, when all effects and side-effects due to the imminence of S on H are accounted for), the joining of H and S in $N = H \vee S$ may still not be 'inverted'. Inversion means to have a set-valued mapping $\Theta : \kappa(N) \;{-}\!\sqsubset\; \kappa(N)$ for which

$$(64) \qquad \Theta \circ \mathrm{Imm}_N = 1_{\kappa(N)}.$$

But as I have explained in Section 4.8, (64) is an equation that the only candidate $\Theta = \mathrm{Imm}_N^{-1}$ would most likely *not* satisfy. Some natural processes are inherently irreversible, and attempts at control can only be from prevention. Attempts at remedy are often circuitous (such as the imminence-driven arguments in this chapter leading to the functorial (63)).

8 Replication

It was our belief then, and I personally still hold to that belief, that in order to develop adequate theories of any complex natural phenomenon, it is necessary at first to investigate *in abstracto* as many as possible conceivable situations. After this, a comparison of data with various abstract situations that have been investigated will show which of the conceivable situations actually may be realized in nature.

—Nicolas Rashevsky (1961)
A Bird's-Eye View of the Development of Mathematical Biology
Proceedings of the Cullowhee Conference on Training in Biomathematics, pp. 11–12

[Replicative] (M,R)-Systems

8.1 Replicare The English action verb 'replicate' comes from the Latin *re + plicare*, 'fold back'. 'Replicate' is, like 'repair', also a word in common usage, and means 'repeat'; 'make a facsimile'; 'reproduce'; 'generate a copy'.

Note that the efficient cause of 'to replicate' is not specified; it can come from without or from within. When an entity replicates itself by its own power or through its inherent nature, then it is termed 'self-replication'.

Viruses cannot self-replicate, but they do replicate by commandeering the reproductive machinery of cells through a process of infection (*RL*: Chapter 13). Cells, given suitable environments, self-replicate by cell division. During cell division, DNA is replicated (with the assistance of intracellular enzymes). Replicated DNA in germ cells gets passed on to offsprings, which may be loosely considered as 'replicates' of the parent organism(s). Prion proteins can replicate by converting normal proteins into rogue forms. The common biological usage of the

A. H. Louie, *Intangible Life*, Anticipation Science 2,
DOI 10.1007/978-3-319-65409-6_9

word 'replicate' is, evidently, not as restrictive as that of the molecule-specific 'repair' in molecular biology.

8.2 Replication in Relational Biology Robert Rosen defined the technical usage of the term *replication* in his (M,R)-systems to mean 'a hierarchicalprocess that entails a repair mapping'. When the repair components themselves need repairing, new mappings that serve to replicate the repair components (or 'repair of repair', 'repair^2') emerge.

Replication is not an *obligatory* feature of (M,R)-*networks*. One may expect that replication is a relatively rare and unusual situation, and most repair mappings are not themselves entailed. *Self*-replication would require that each replication be entailed from the metabolism and repair components in the network. Such self-sufficiency depends on some "stringent but not prohibitively strong conditions" imposed on the mappings involved. Recall (Definition 7.7) that an (M,R)-*system* is an (M,R)-network that is closed to efficient causation. The conditions for 'entailment closure' will not usually be satisfied, whence most (M,R)-networks are not (M,R)-systems.

In Rosen's early writings, he used the term '(M,R)-system' to mean a network of metabolism and repair components in which metabolism $\vdash$ repair, but the network did not necessarily satisfy the stringent requirements for entailment closure. Since an '(M,R)-system' (in this old sense) did not usually have all its repair components replicated within, one that did so acquired an adjective and was called a *replicative (M,R)-system*. The defining characteristic is the self-sufficiency in the networks of metabolism–repair–replication components, in the sense that every mapping is entailed within; in short, *closure to efficient causation* ('*clef*'). In all his later publications (notably including the definitive survey Rosen [1972]), however, the adjective 'replicative' was routinely omitted and an '(M,R)-system' verily denoted one that was clef.

To elucidate matters, I have coined the term *(M,R)-network* (*ML*: 11.10, 11.13, 13.2) to describe a network of metabolism and repair components that is not necessarily closed to efficient causation. I have also dropped the adjective 'replicative' for (M,R)-systems, whence *all (M,R)-systems are replicative*. Thenceforth these are the senses of the terms consistently used in relational biology. These definitions were repeated in 7.6 and 7.7 for completeness. Notably, it is in the clef sense that the term '(M,R)-system' appears in the Postulate of Life (7.8), that *a natural system is an organism if and only if it realizes an (M,R)-system*.

8.3 Replication is Special Repair Metabolism is material entailment, which means it is a process $f : a \mapsto b$ with a final cause (output) b that is used as material cause (input) of another process, say $g : b \mapsto c$. Its relational diagram in graphic-theoretic form is that of sequential composition:

(1)

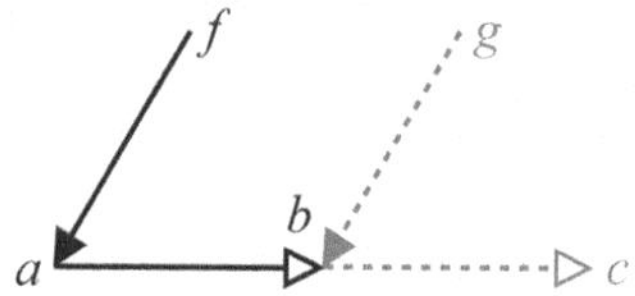

In diagram (1), I have shown the arrows of metabolism $f : a \mapsto b$ in black with solid lines, and the immanently entailed sequential relay $g : b \mapsto c$ in grey with dashed lines.

Repair is functional entailment, which means it is a process $\Phi : x \mapsto f$ with a final cause f that is itself the efficient cause of another process, say $f : a \mapsto b$. Its relational diagram in graphic-theoretic form is that of hierarchical composition:

(2)

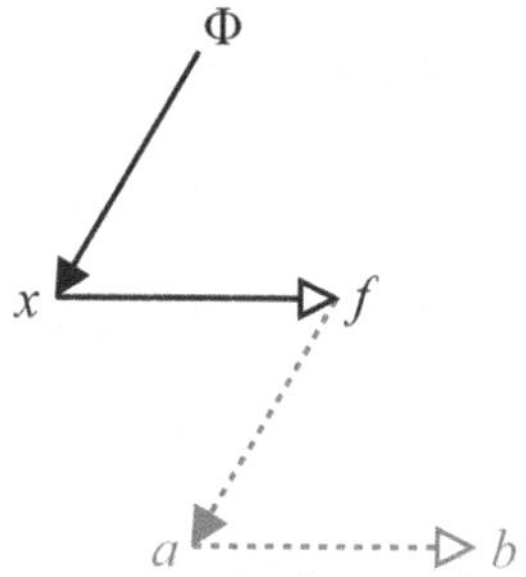

In diagram (2), the arrows of repair $\Phi : x \mapsto f$, the subject, are likewise in black with solid lines, and the immanently entailed hierarchical relay $f : a \mapsto b$ in grey with dashed lines.

One readily observes from the relational diagrams (1) and (2) that metabolism and repair are different in kind. The topologies of their respective networked arrows are distinct: diagram (1) contains a branching at node b while diagram (2) is a simple chain. In terms of the collection $\kappa(N)$ of processes in an (M,R)-network $\langle N, \kappa(N) \rangle$, the set M of metabolism processes and the set R of repair processes are distinct (although not necessarily mutually exclusive; it may happen that $\mathrm{M} \cap \mathrm{R} \neq \varnothing$, as explicated in Sections 7.11 and 7.12).

Replication is also functional entailment. It is a process $\beta : y \mapsto \Phi$ with a final cause Φ that is the efficient cause of another process, $\Phi : x \mapsto f$. But, in addition, this entailed process Φ is *itself* functional entailment. Its relational diagram in graphic-theoretic form is that of a hierarchical chain of three processes:

(3)

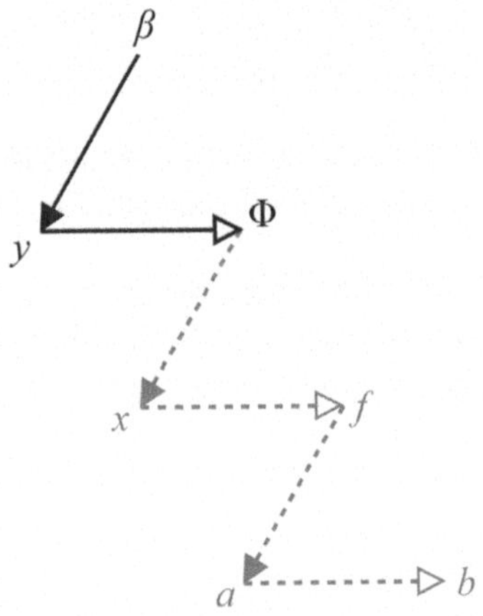

(again showing the subject in solid black and the immanently entailed processes in dashed grey). The relational diagrams (2) and (3) of repair and replication are different in degree but not in kind. Diagram (2) is a hierarchical chain of two mappings while diagram (3) is a hierarchical chain of three mappings.

In summary, replication is a special repair process that happens to be repair^2. But, in terms of the collection $\kappa(N)$ of processes in an (M,R)-network $\langle N, \kappa(N) \rangle$, the set of replication processes forms a subset of the set of repair processes, which one may tersely denote $\mathrm{R}^2 \subset \mathrm{R}$. This containment is the reason that the networks and systems are called *metabolism–repair*, not 'metabolism–repair–replication'.

Genesis of Replication

8.4 Domains of Replication In the simplest (M,R)-network

(4)
$$A \xrightarrow{f} B \xrightarrow{\Phi} H(A,B),$$

a replication map must have as its codomain the hom-set $H(B,H(A,B))$ to which repair mappings Φ belong. So it must be of the form

(5)
$$\beta : Y \to H(B,H(A,B))$$

for some set Y.

The most important feature of an (M,R)-system is the closure of its entailment structure. To make the network (4) clef, Y must be a set already present in the form. So one may have

(6) $$Y = H(A, B),\ B,\ \text{or}\ A$$

In the multi-component general (M,R)-network (Definition 7.6), the choices, independently varying for each index $i \in I$, would be $Y = H(A_j, B_j)$, $\prod_{k=1}^{n} B_{i_k}$, or A_k.

That for an (M,R)-system the replication map (i.e., that which entails the repair map Φ) must already be entailed in the form (4) is equivalent to the graph-theoretic requirement that digraph

(7)

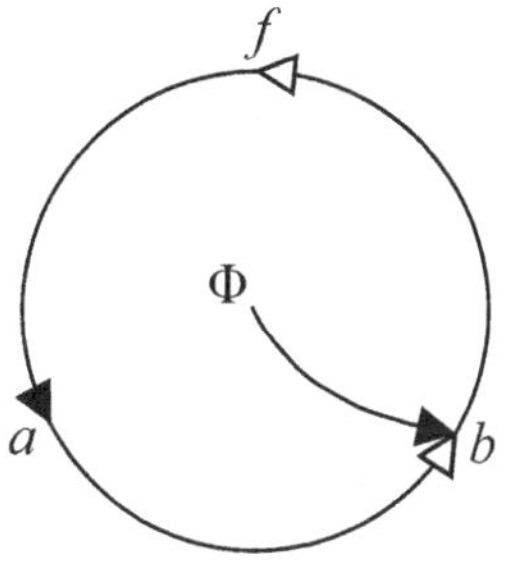

be completed so that it has a closed path containing all the solid-headed arrows.

8.5 First Class of (M,R)-Systems When $Y = H(A, B)$, which has been Rosen's choice for replication in all his publications, one has $\beta : f \mapsto \Phi$. This corresponds to the relational diagram

(8)

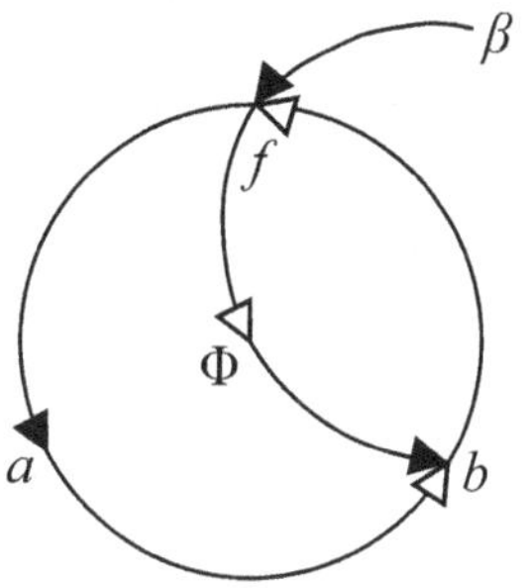

In order to entail the replication map

(9) $$\beta : H(A,B) \to H(B,H(A,B))$$

within, it may be constructed as the *inverse evaluation map*

(10) $$\beta = \hat{b}^{-1} : f \mapsto \Phi$$

(*ML*: 11.15–11.23, 12.2). Since $b \in B$ uniquely determines $\hat{b}^{-1} \in H(H(A,B), H(B,H(A,B)))$, a bijection

(11) $$b \cong \hat{b}^{-1}$$

is established (between B and a subset of $H(H(A,B),H(B,H(A,B)))$ containing all inverse evaluation maps $\beta = (\hat{\bullet})^{-1}$). This first kind of replication may be represented as

(12) $$b : f \mapsto \Phi.$$

Thus the relational diagram becomes clef, and yields the simplest (M,R)-system:

(13)

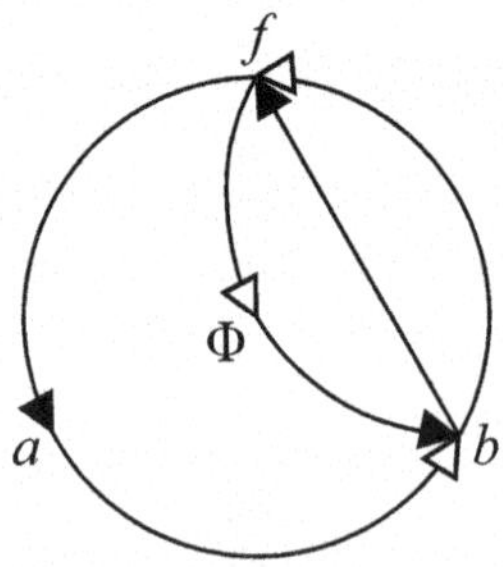

8.6 Second Class of (M,R)-Systems The alternative $Y = B$ leads to the second kind of replication, with $\beta : b \mapsto \Phi$. This corresponds to the relational diagram

(14)

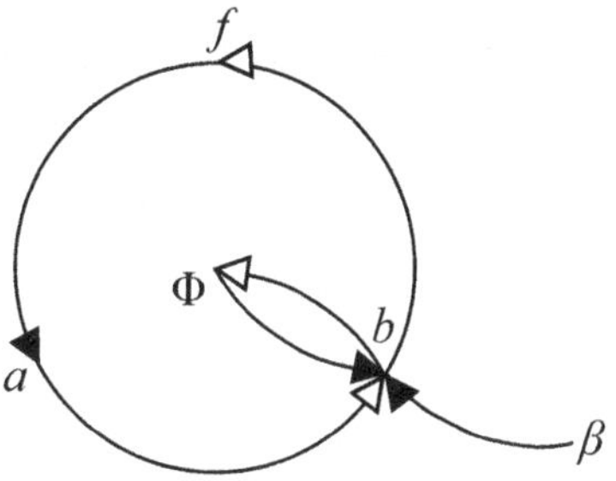

The replication map

(15) $$\beta : B \to H(B, H(A, B))$$

may be considered as the abstract version of the *conjugate isomorphism* of a Hilbert space onto its dual space (*ML*: 12.3–12.9),

(16) $$\beta = \gamma_B : b \mapsto \Phi.$$

It turns out that $b \in B$ likewise uniquely determines $\gamma_B \in H(B, H(B, H(A, B)))$. So an isomorphism

(17) $$b \cong \gamma_B$$

is also established, and this second kind of replication may be represented as

(18) $$b : b \mapsto \Phi,$$

with b serving as both the efficient cause (albeit in an isomorphic reincarnation) and the material cause. The relational diagram is then clef:

(19)

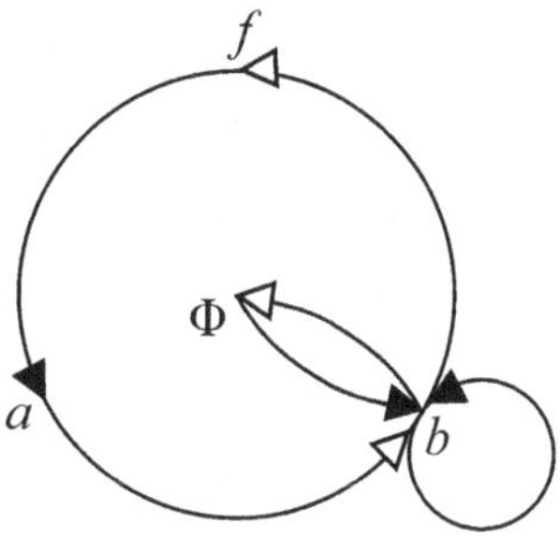

8.7 Third Class of (M,R)-Systems The third and final choice for the domain of the replication map is $Y = A$. Now the replication must be $\beta : a \mapsto \Phi$:

(20)

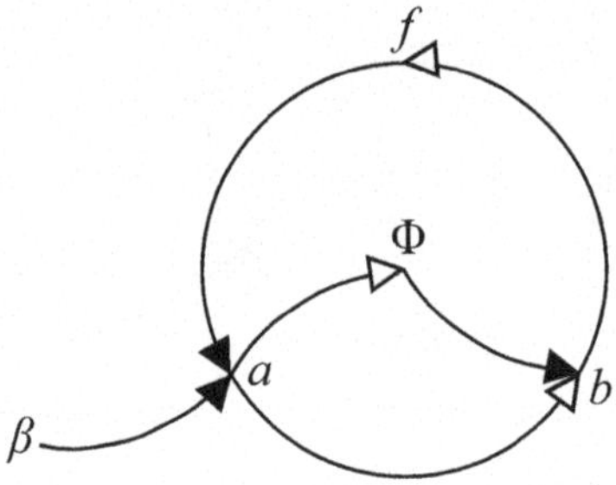

The replication map

(21) $$\beta : A \to H(B, H(A, B))$$

may be considered as the generalized natural projection (*ML*: 12.10–12.12)

(22) $$\beta = \pi_S \circ (\cdot)^{-1} : a \mapsto \Phi.$$

With this correspondence, an $a \in A$ defines the replication map Φ to be the similarity class of a^{-1}. The isomorphism is thus

(23) $$a \cong \pi_S \circ (\cdot)^{-1},$$

and this third kind of replication may be represented as

(24) $$a : a \mapsto \Phi,$$

this time with a serving as both the efficient cause (through a convoluted isomorphism) and the material cause. The relational diagram is hence clef:

(25)

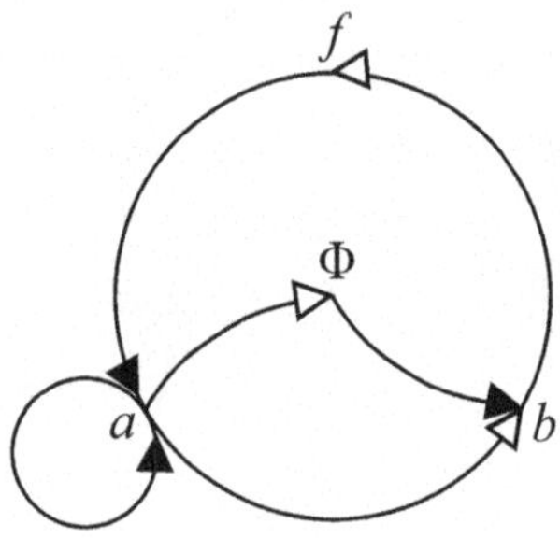

Repair[2] of the First Kind

8.8 Hom-Set of the Monomorphism $b \mapsto \hat{b}^{-1}$ The inverse evaluation map (10) that defines the first kind of replication map establishes the correspondence $b \cong \hat{b}^{-1}$ (11), which is the monomorphism (*ML*: A.43)

$$\text{(26)} \qquad \mathrm{M}_1 : B \to H(H(A,B),H(B,H(A,B)));$$

i.e.,

$$\text{(27)} \qquad \mathrm{M}_1 \in H(B,H(H(A,B),H(B,H(A,B)))).$$

The transformation $\mathrm{M}_1 : b \mapsto \hat{b}^{-1}$ verily offers a model of *bioactivity*: it is through the monomorphic correspondence M_1 that a tangible material product b acquires its intangible potential as an efficient cause $\hat{b}^{-1}$.

8.9 Set-Isomorphism **Set** is a cartesian closed category (*cf. ML*: A.53 and A.19 (iii)), and therein one has the following isomorphism:

$$\text{(28)} \qquad H(X \times Y,Z) \cong H(X,H(Y,Z))$$

(cf. Section 0.26).

To understand the isomorphism (28) in **Set**, one observes that a mapping of two arguments, $g(x,y) = z$, may be evaluated sequentially, one argument at a time, as $g(x,\bullet)(y) = z$. This natural equivalence transforms the material entailment

$$\text{(29)} \qquad g : (x,y) \mapsto z$$

into the functional entailment

$$\text{(30)} \qquad g^{\rightarrow} : x \mapsto g(x,\bullet) \quad \text{with} \quad g(x,\bullet) : y \mapsto z.$$

On the other hand (which in category-theoretic terms is 'as the right adjoint', a subject that I shall explore in detail in *IL*: Part III), a hierarchical composition

$$\text{(31)} \qquad h : x \mapsto k \quad \text{with} \quad k : y \mapsto z$$

may be compressed into a single mapping of two arguments by 'evaluation at y', as

(32) $$h^{\leftarrow} : (x,y) \mapsto h(x)(y) = k(y) = z.$$

8.10 Repair of Repair When the **Set**-isomorphism (28) is applied to the hom-set $H(B, H(H(A,B), H(B, H(A,B))))$ of (27), one obtains the bijection

(33) $$\begin{aligned} &H(B \times H(A,B), H(B, H(A,B))) \\ &\cong H(B, H(H(A,B), H(B, H(A,B)))) \end{aligned}.$$

Note the form of the hom-set on the left-hand side. Recall (Definition 2.1) that a mapping is a special subset of the cartesian product (viz. a mapping $g : X \to Y$ is a special relation between sets X and Y, a subset $g \subset X \times Y$ with a unique-value property, such that if $(x, y_1) \in g$ and $(x, y_2) \in g$ then $y_1 = y_2$). So $H(X,Y) \subset \mathrm{P}(X \times Y)$. In particular,

(34) $$H(B, H(A,B)) \subset \mathrm{P}(B \times H(A,B))$$

(cf. Remarks 7.14), and $H(B, H(A,B))$ is the hom-set to which the repair map Φ belongs!

Indeed, the hom-set on the left-hand side of (33) contains two copies of Φ. The domain $B \times H(A,B)$ of this hom-set contains $\Phi \subset B \times H(A,B)$ as a relation, in the form of

(35) $$\Phi = \{(b,f) \in B \times H(A,B) : \Phi(b) = f\};$$

the codomain $H(B, H(A,B))$ of this hom-set contains $\Phi \in B \to H(A,B)$ as a mapping. In other words, the monomorphism $\mathrm{M}_1 : b \mapsto \hat{b}^{-1}$, in its equivalent form as

(36) $$\mathrm{M}_1^{\leftarrow} : (b,f) \mapsto \mathrm{M}_1(b)(f) = \hat{b}^{-1}(f) = \Phi,$$

offers up a self-referential double-dose of Φ.

This means, a fortiori, that the action sequences explicated in Sections 7.15 and 7.16 may also be considered a biological realization of the isomorphism

$b \cong \hat{b}^{-1}$, in addition to that of the repair map $\Phi : \{b\} \mapsto [\![b]\!] \mapsto f$. The monomorphism $\mathrm{M}_1 : b \mapsto \hat{b}^{-1}$ is, of course, what defines the first kind of replication map (12):

(37)

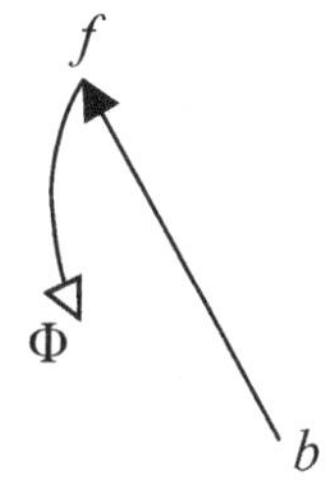

The equivalent monomorphism $\mathrm{M}_1^{\leftarrow} : (b,f) \mapsto \Phi$ that is the iteration $\langle \Phi, \Phi \rangle$ perfectly illustrates the role of replication as 'repair of repair'.

Ouroboros

8.11 Ουροβόρος The *ouroboros*, the 'tail-devouring snake', is an ancient symbol depicting a serpent (or dragon) swallowing its own tail and forming a cycle. It is often used as a metaphor for *self-referencing*.

(38)

This drawing (adapted from the *Codex Marcianus graecus*, c.10th/11th century) is a reproduction from the early (2nd-century Alexandrian) alchemical text *The Chrysopoeia of Cleopatra*. The enclosed Ancient Greek words are ἕν ('one') τό ('the') πάνοψ (written through contraction, 'all seeing/including'), whence "One is All" (which by symmetry of the binary relation 'is', i.e., '=', may also be interpreted as "All is One"). This slogan means that all substances in nature are one so that an alchemist's purpose is to find a method of unifying various separate materials (and, say, recombining them to make gold).

8.12 Ouroboros Mappings In *ML*: 5.12, I have considered the possible and impossible ouroboros for a mapping

(39) 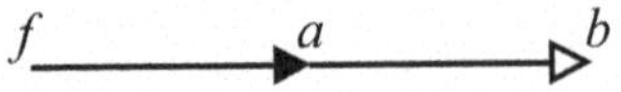

Among the three vertices $\{f, a, b\}$ of the digraph, when $a \cong b$ one has the *self-inference*

(40) 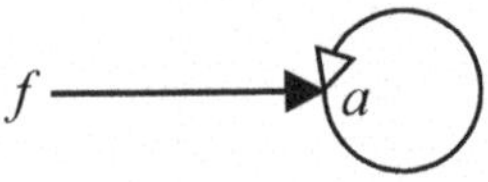

i.e., $f : a \mapsto a$. This is an *automorphism* that may represent either the *identity mapping* $1_A \in H(A, A)$ or the *fixed point a* of the mapping $f : A \to A$.

When $b \cong f$, the *self-entailed mapping* that results,

(41) 

i.e., $f : a \mapsto f$, is an impossibility in the category **Set**, except trivially when its domain A is either empty or a singleton set. This is because the existence of '$f \vdash f$' (equivalently, for a mapping to be contained in its own range, $f \in \operatorname{ran}(f)$) would involve an infinite hierarchy of hom-sets:

(42) $$f \in H(A, H(A, H(A, H(A, \cdots)))).$$

Vacuously, for an empty domain, since $H(\varnothing, Y) = \{\varnothing\}$ for any set Y (*cf.* Section 2.4), the hierarchy of hom-sets in (42) collapses to $\{\varnothing\}$. Thus f is the empty mapping $\varnothing$, whence the tautology $\varnothing \vdash \varnothing$ ("*Ex nihilo nihil fit.*", as it were). If A is a singleton set, then f is clearly determined by its *only* functional value.

Given a mapping f, one of course has f itself. But this is the statement of the entailment of existence '$f \vdash \exists f$' (see *ML*: 5.18 on immanent causation). The ouroboros '$f \vdash f$' requires a causation different in kind. Note also that the impossibility of the existence of the nontrivial ouroboros $f \vdash f$ is a statement in naive set theory (and for categories of sets with structure, which form the universe with which we are concerned). One may, however, expand the universe (and one's horizons) at will in mathematics. In hyperset theory, for example, $f \vdash f$ does exist, and is precisely analogous to the prototypical hyperset equation $\Omega = \{\Omega\}$, which has a unique solution.

8.13 Self-Reproduction The impossibility of $f \in \mathrm{ran}(f)$ in **Set** was, incidentally, the subject of a couple of early Rosen publications [Rosen 1959, 1962], in the context of self-reproducing automata. Indeed, the logical paradox arising from the definition of 'self-reproduction' as $f \vdash f$ was, perhaps as an inside joke, referred to as the "Rosen paradox" by Nicolas Rashevsky [Rashevsky 1961]. The fact that $f \notin \mathrm{ran}(f)$ (whence $f \notin \mathrm{Imm}(f)$; cf. Definition 6.12) and, in the terminology introduced in *RL*, and that every $f \in \mathfrak{R}\mathbf{C}$ is *independent of itself*, have important consequences in relational biology (cf. *RL*: 7.15 et seq.).

The resolution of the Rosen paradox was achieved by modifying the definition of 'self-reproduction' (for an automaton in the original exercise) so that it is no longer the stringent requirement of '$f \vdash f$'. The relation diagram (13) of the simplest (M,R)-system (with replication of the first kind) serves as an illustration. Therein the (M,R)-system maps of metabolism–repair–replication are

(43) $$\{f : a \mapsto b, \Phi : b \mapsto f, b : f \mapsto \Phi\}.$$

The entailment pattern that is a hierarchical cycle of these three maps is more evident when the relational diagram is unfolded thus:

(44)

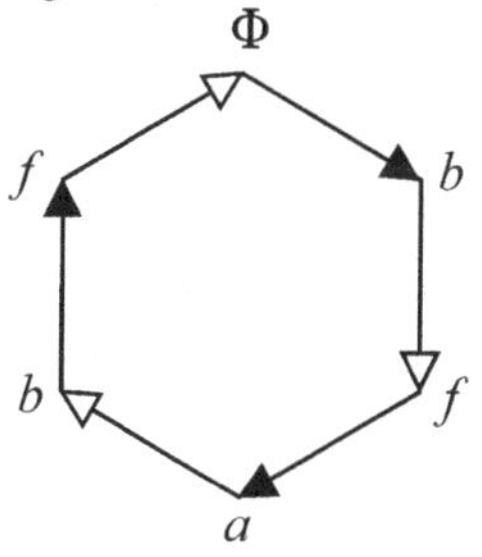

One may also note that there is no 'privileged' position of any of the three mappings involved. They may be assigned the labels of metabolism, repair, and replication in any cyclic permutation.

This cyclic entailment is the true realization of the ouroboros axiom έν τό πάνοψ. The all-important feature is that the mappings form a hierarchical cycle. Stated otherwise, it is the hierarchical-cycle that makes the (M,R)-system a model of a (self-reproducing) cell. I have shown in Section 7.12 that the subgraph (7) of (13) shows $f : a \mapsto b$ and $\Phi : b \mapsto f$ in both hierarchical and sequential composition, and the subgraph

(45)

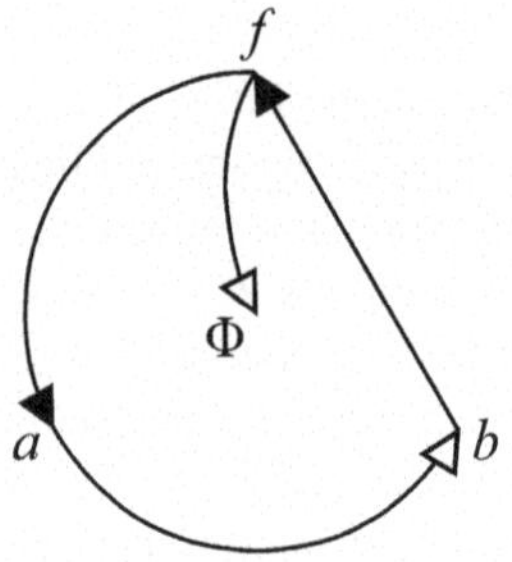

shows $f : a \mapsto b$ and $b : f \mapsto \Phi$ in hierarchical composition. The final pairing, that of $\Phi : b \mapsto f$ and $b : f \mapsto \Phi$, gives the subgraph

(46)

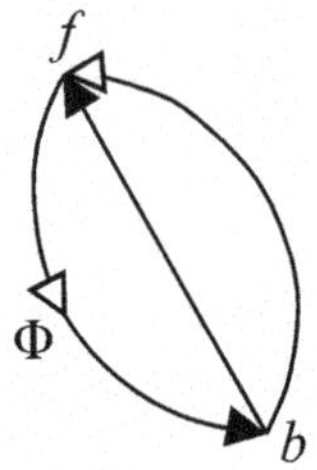

It is through the combination of the two functional entailments $\Phi \vdash f$ and $b \vdash \Phi$ that one obtains the indirect 'self-reproduction $\Phi \vdash \Phi$'. In other words, while a self-entailed mapping in **Set** is impossible, relational diagram (46) suffices to serve as the arrow diagram of self-entailment if one slightly relaxes the definition.

Repair² of the Second and Third Kinds

8.14 Self-Referencing Processor The final type of ouroboros mapping, when $f \cong a$ among the three vertices $\{f, a, b\}$ of the digraph (39), is illustrated in the relational diagram

(47)

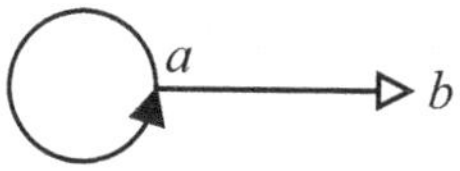

Note that both the second and third kind of replication maps, as represented in relational diagrams (19) and (25), embody this pattern.

Arrow diagram (47) contains the *self-referencing processor*

(48) $$a \longrightarrow a$$

The 'self-referencing' symbolism does not mean that instead of $f(a) = b$ one has '$a(a) = b$'. The situation represented herein is, rather, where a mapping $f : A \to B$ is uniquely determined by a specific element $a \in A$ in its domain. As a simple example, for a *fixed* $a_0 \in A$, consider the mapping $f : A \to \{0, 1\}$ defined by $f(a_0) = 1$, and $f(a) = 0$ for $a \neq a_0$; such f is, indeed, the *characteristic mapping* $\chi_{\{a_0\}}$ (Definition 1.13). *Each* $a_0 \in A$ determines its corresponding characteristic mapping $\chi_{\{a_0\}}$ uniquely. The identification $a \leftrightarrow \chi_{\{a\}}$ establishes a correspondence (i.e., an isomorphism) between A and a subset of the set $H(A, \{0, 1\}) \cong 2^A$ of all morphisms from A to $\{0, 1\}$. Isomorphic objects are considered categorically the same, thus, for this example, arrow diagrams (47) and (48) are abstract representations of '$a \cong \chi_{\{a\}} : A \to \{0, 1\}$'.

In relational diagram (19), the $b : b \mapsto \Phi$ represents the conjugate isomorphism $b \cong \gamma_B : b \mapsto \Phi$, and in relational diagram (25), the $a : a \mapsto \Phi$ denotes the similarity class of the inverse $a \cong \pi_S \circ (\cdot)^{-1} : a \mapsto \Phi$.

8.15 Enzyme Specificity The second kind of replication map establishes the isomorphism $b \cong \gamma_B$ (17), which is the monomorphism

(49) $$\mathrm{M}_2 : B \to H(B, H(B, H(A,B)));$$

i.e.,

(50) $$\mathrm{M}_2 \in H(B, H(B, H(B, H(A,B)))).$$

The isomorphism (28), when applied to the hom-set in (50), gives

(51) $$H(B \times B, H(B, H(A,B))) \cong H(B, H(B, H(B, H(A,B)))).$$

The hom-set on the left-hand side of (51) is such that its codomain $H(B, H(A,B))$ contains the repair map $\Phi \in B \to H(A,B)$, while its domain $B \times B$ is precisely the domain of the 'generalized inner product' $\langle \cdot, \cdot \rangle : B \times B \to H(A,B)$ used in the entailment of the repair map $\Lambda_b = \gamma_B(b) = \Phi \in H(B, H(A,B))$ via

(52) $$\Phi(x) = \Lambda_b(x) = \langle x, b \rangle \in H(A,B) \quad \text{for } b, x \in B$$

(details in *ML*: 12.7).

The relational diagram (19) unfolds into

(53)

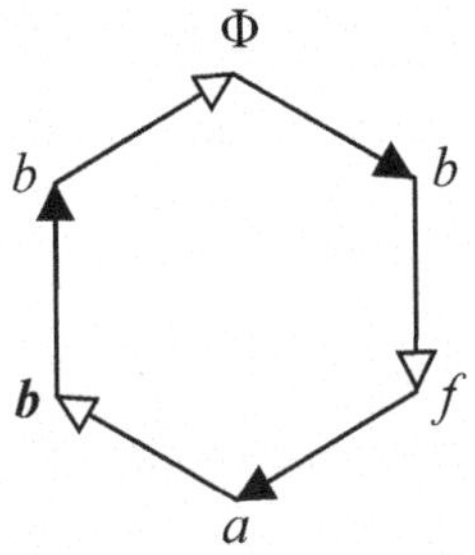

The noteworthy feature here is that there are three vertices 'b'. The top two b s are material causes; the bold $\boldsymbol{b}$ vertex (the relay node in the hierarchical composition $f \vdash \boldsymbol{b} \vdash \Phi$) is final cause in $f : a \mapsto \boldsymbol{b}$ and efficient cause in $\boldsymbol{b} : b \mapsto \Phi$. The 'double dose' of b in $\boldsymbol{b} : b \mapsto \Phi$ is the embodiment of the monomorphism

(54) $$\mathrm{M}_2^{\leftarrow} : (\boldsymbol{b}, b) \mapsto \mathrm{I}_2(\boldsymbol{b})(b) = \gamma_{\boldsymbol{B}}(b) = \Phi.$$

As explained in *ML*: 12.8, the requirement that a repair map Φ [gene] be uniquely determined by a metabolic product $b \in B$ in its domain may be realized thus. A metabolic product in fact determines the enzyme $f \in H(A,B)$ required in the biochemical reaction that produces it. This is the concept of *enzyme specificity*. The one-gene-one-enzyme hypothesis then completes the entailment path to the gene $\Phi \in H(B,H(A,B))$.

8.16 Protein Biochemistry The third kind of replication map establishes the isomorphism $a \cong \pi_S \circ (\cdot)^{-1}$ (23), which is the monomorphism

(55) $$\mathrm{M}_3 : A \to H(A,H(B,H(A,B)));$$

i.e.,

(56) $$\mathrm{M}_3 \in H(A,H(A,H(B,H(A,B)))).$$

The isomorphism (28), when applied to the hom-set of (56), gives

(57) $$H(A \times A, H(B,H(A,B))) \cong H(A,H(A,H(B,H(A,B)))).$$

The hom-set on the left-hand side of (57) is such that, as in the case for the other two kinds of replication, its codomain $H(B,H(A,B))$ contains the repair map $\Phi \in B \to H(A,B)$ (which is, after all, the *raison d'être* of replication). The domain $A \times A$ now contains the similarity class of the inverse

(58) $$\beta : a \mapsto \pi_S(a^{-1}) = [fa^{-1}f^{-1}]_S = [a^{-1}]_S = \Phi,$$

the mathematical details of which are explained in *ML*: 12.12. The relational diagram (25) unfolds into

(59)

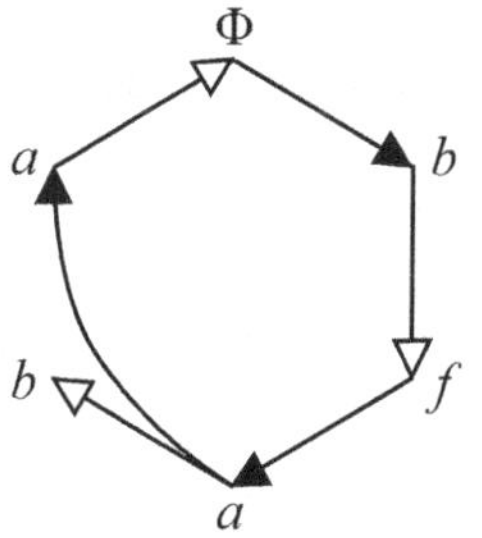

The geometry of this graph is slightly more complicated than the simple cycles of (44) and (53), and has been explicated in *ML*: 12.16.

The equivalent monomorphism is

(60) $$\mathrm{M}_3^{\leftarrow} : (a,a) \mapsto \mathrm{I}_3(a)(a) = \pi_S(a^{-1}) = \Phi.$$

As also noted in *ML*: 12.12, the final cause, that a repair map is identified with an equivalence class of substrates, may be realized thus. The concept of enzyme specificity applies just as well from substrates $a \in A$ to enzymes $f \in H(A,B)$ required in the biochemical reactions that metabolize them. The one-gene-one-enzyme hypothesis then, again, completes the entailment path to the gene $\Phi \in H(B,H(A,B))$. Also, because of the self-referencing nature of the replication map $\beta : a \mapsto [a^{-1}]_S$, the set of metabolism-repair-replication maps for this class of (M,R)-system may be decoded into the set of pathways of *protein biochemistry*. In particular, the enzymes involved act on enzymes themselves, and may be realized among peptide synthases, protein polymerases, protein kinases, and peptidases.

Έν τό πάνοψ

8.17 Clef The monomorphisms $\mathrm{M}_1 : b \mapsto \hat{b}^{-1}$, $\mathrm{M}_2 : b \mapsto \gamma_B$, and $\mathrm{M}_3 : a \mapsto \pi_S \circ (\cdot)^{-1}$ provide three models of transformations through which material causes become efficient causes, how structures acquire functions. They are manifested in (M,R)-systems as three kinds of replication maps, but the concepts are general, and through

(61) $$\text{replication} = \text{repair}^2 = [\text{functional entailment}]^2$$

these processes fall entirely within the auspices of the imminence mapping $\mathrm{Imm}_{\mathbf{C}}$ of the category **C** of (M,R)-systems.

The naturally equivalent monomorphisms that define the three kinds of replication are $\mathrm{M}_1^{\leftarrow} : (b,f) \mapsto \Phi$, $\mathrm{M}_2^{\leftarrow} : (\boldsymbol{b},b) \mapsto \Phi$, and $\mathrm{M}_3^{\leftarrow} : (a,a) \mapsto \Phi$. All three functionally entail in turn the hierarchical entailment 'repair ⊢ metabolism'; that is, $\Phi \vdash f$ (whence the action chains explicated in Sections 7.15 and 7.16 on the nuances of repair action). That the three kinds of replication have different efficient causes serve to emphasize the nuances of replication action under different circumstances. But all three have a morphological and functional unity, which may be summarized in this one symbol, the entailment cycle that is relational biology's own ouroboros:

(62)

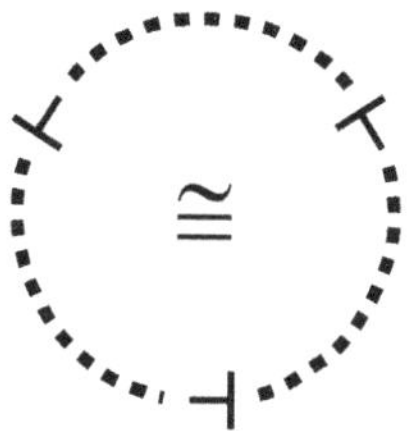

8.18 Anticipation Anticipation is a necessary condition of life: a living system anticipates. Indeed, an (M,R)-system is anticipatory, and an anticipatory system is an (M,R)-network. Self-replication is the quintessential anticipatory activity of a cell. Closure in efficient causation implies a continuous fabrication of internal processes in anticipation of the necessity of future repairs.

IL is, however, not the forum in which to discuss these connections. I refer the interested reader to the Springer *Handbook of Anticipation*, in particular the chapters 'Basic Biological Anticipation' [Hofmeyr 2017], 'Complex Systems' [Louie & Poli 2017], 'Relational Biology' [Louie 2017a], and 'Mathematical Foundations of Anticipatory Systems' [Louie 2017b].

Part III
Dimissio
From Invertibility to Adjunction

—Ioannes Petrus Aloisius Praenestinus
[Giovanni Pierluigi da Palestrina] (c. 1525–1594)
Pater noster

Descent

The encoding functor $\varepsilon : \langle N, \kappa(N) \rangle \rightarrow \langle M, \kappa(M) \rangle$ is a very elaborately constructed sample of a mapping, but a mapping nonetheless. A mapping forgets therefore an encoding functor forgets. A large part of the art of modelling is, despite the inevitable loss of information inherent in the encoding, how one may invert the entropic effects and construct a decoding to optimize the recovery of information. Invertibility, however, is a matter of degree.

In Part III, I shall explore functorial 'inverses' in various senses, with special emphasis on what they entail in the modelling relation. In particular, inverses of the power set functor descend hierarchical levels, and again serve to connect and integrate the join of parts $\vee P$ and the whole W. Level-transcending functors compose and iterate the analyses into parts and syntheses of wholes. The functorial connections among alternate descriptions, in the absence of a most-favoured level, are the hallmark of complex systems.

9
Equivalence

Die bürgerliche Gesellschaft ist beherrscht vom Äquivalent. Sie macht Ungleichnamiges komparabel, indem sie es auf abstrakte Größen reduziert. Der Aufklärung wird zum Schein, was in Zahlen, zuletzt in der Eins, nicht aufgeht; der moderne Positivismus verweist es in die Dichtung.

[Bourgeois society is ruled by equivalence. It makes dissimilar things comparable by reducing them to abstract quantities. As for Enlightenment, that which does not reduce to numbers, and ultimately to the one, becomes illusion; modern positivism consigns it to poetry.]

—Max Horkheimer and Theodor W. Adorno (1947)
Dialektik der Aufklärung [*Dialectic of Enlightenment*]
Begriff der Aufklärung

Mappings Lose Information

The prefect model of a set is the set itself, possibly embedded in a superset (Definition 1.1). For the embedding $A \subset X$, the single-valued mapping $i : A \to X$ defined by $i(x) = x$ [also, the set-valued mapping $i : A \multimap X$ defined by $i(x) = \{x\}$], for all $x \in A$, is called the *inclusion map* of A in X. The inclusion map of X in X is called the *identity map* on X, denoted 1_X (Definition 1.2). As relations, the inclusion map is the set $i = \{(x,x) : x \in A\} \subset A \times X (\subset X \times X)$, and the identity map is the set $1_X = \{(x,x) : x \in X\} \subset X \times X$. Thus each is a member of $\mathrm{P}(X \times X)$ that consists of all the diagonal elements corresponding to the embedded set (cf. Section 1.25).

A. H. Louie, *Intangible Life*, Anticipation Science 2,
DOI 10.1007/978-3-319-65409-6_10

9.1 A Mapping Models One function of a mapping $f : X \rightarrow\!\sqsubset Y$ is to compare its domain X with its codomain Y. The comparison endows f with the role of an encoding: the range $\operatorname{ran}(f) = f(X)$ may be considered a model of the domain X, contained in a larger model that is the codomain Y. That is, $f(X) \leq Y$ in the lattice $\mathbf{C}(X)$ of models of X (*ML*: 7.28). One may naturally ask whether distinct features of X remain distinguishable in its model $f(X) \in \mathbf{C}(X)$. If not, then some informational content of X has been lost in its image $f(X) \subset Y$. This loss, congenital to the very definition of mapping, is the manifestation of nature's entropy in mathematics.

9.2 Injection It is only in the rare case of *injective* (single-valued) mappings $f : X \to Y$ (Definition 2.11) that one has the distinctiveness-preservation property

(1) $$x_1 \neq x_2 \quad \Rightarrow \quad f(x_1) \neq f(x_2)$$

(*ML*: 1.8, *RL*: 1.13), which is equivalent contrapositively to

(2) $$f(x_1) = f(x_2) \quad \Rightarrow \quad x_1 = x_2.$$

Intuitively, an injection maps the set X in a one-to-one fashion onto the range $f(X) \subset Y$. Stated otherwise, the codomain Y of an injective mapping contains a *lossless* model $f(X)$ of the domain X.

9.3 A Mapping Forgets A mapping is not necessarily injective (Section 2.27), and therefore loses information. A non-injective mapping f 'forgets' the distinctiveness of some members of its domain, by identifying $f(x_1) = f(x_2)$ when $x_1 \neq x_2$. There are many ways to forget a multitude of things. A *forgetful functor* (Definition 0.16) from a category $\mathbf{C}$, of 'sets with structure' and homomorphisms that preserve this structure, to **Set** assigns to each $\mathbf{C}$-object its underlying set, and regards each homomorphism as a mapping of sets. (I shall revisit forgetful functors later.)

9.4 A Model Forgets That injectivity of $f : X \to Y$ implies $|X| \leq |Y|$ is the Pigeonhole Principle (Axiom 2.19). A lossless model defined by an injection therefore requires a *larger* codomain, while modelling is usually used as a tool with which one studies simpler, easier-to-analyse, hence *smaller*-than-the-original, systems: one sees the opposing tensions. And opposing tensions argue for *optimization.* Goldilocks beckons.

Modelling is the art of bringing entailment structures into congruence. The selection is on the degree of incompleteness, on what entailments to include and what to leave out. The choice of *le modèle juste* amounts to a balance between William of Ockham's principle of parsimony "entia non sunt multiplicanda praetor necessitatem" [entities must not be multiplied beyond necessity] and Roberto Poli's witty rejoinder "entia non sunt diminuenda sine necessitate" [entities must not be diminished without necessity]. Therein lies the art.

Encoding, by its very nature, is analytic; in breaking the whole into parts something is inevitably lost. Decoding, the craft of undoing, on the other hand, is synthetic; the craft is to enlist the intangible and optimally put Humpty Dumpty together again. The art of modelling is in the decoding.

Invertibility and Injectivity

Injectivity implies invertibility. A non-injective mapping 'forgets'. But how much can one loosen the property of injectvity or extend the notion of invertibility, and still retain a satisfactory amount of fidelity? The exposition of this optimization problem is *IL*'s final cause; the efficient cause is the assorted functorial connections.

9.5 Definition If two (single-valued) mappings $f : X \to Y$ and $g : Y \to X$ compose to the identity,

$$g \circ f = 1_X, \tag{3}$$

then the mapping g is called a *left inverse* of f, while f is called a *right inverse* of g.

Consider the "take the positive square root" mapping $f : X \to Y$ from $X = \{x \in \mathbb{R} : x \geq 0\}$ to $Y = \mathbb{R}, f(x) = \sqrt{x}$. The equation $(\sqrt{x})^2 = x$ means the "square" mapping $g : Y \to X$, $g(y) = y^2$, is the left inverse of f. But $f \circ g \neq 1_Y$ (e.g., $\sqrt{(-2)^2} = 2 \neq -2$), so g is not the right inverse of f. This simple example illustrates that the left–right roles of inverses are not in general interchangeable.

Recall Lemma 2.14, that if $g \circ f = 1_X$ then f is injective and g is surjective. Thus a left inverse is surjective, and a right inverse is injective. Indeed, one has the

9.6 Theorem *For $X \neq \varnothing$, a (single-valued) mapping $f : X \to Y$ is an injection if and only it has a left inverse $g : Y \to X$.*

The caveat $X \neq \varnothing$ is for the bypassing of the empty-set pathology. This is because the empty mapping $\varnothing : \varnothing \to Y$ is injective; but when $Y \neq \varnothing$, $\varnothing^Y = \varnothing$ (cf. Sections 2.4 and 2.11), so there cannot be any mapping $g : Y \to \varnothing$ to fulfill the role of the left inverse in $g \circ \varnothing = 1_\varnothing (= \varnothing)$.

Constructively (when $X \neq \varnothing$), the left inverse g takes each element of the image $f(X)$ back to the necessarily unique element of X from which it came, and the remaining elements of Y (i.e., those elements in $Y \sim f(X)$, if any) may each be sent to any element of X. Note that members of $Y \sim f(X)$ are not involved in the compositional relay $g \circ f : x \mapsto f(x) \mapsto x$, so g may be arbitrarily defined there. This freedom of choice on $Y \sim f(X)$ explains why a left inverse is not necessarily uniquely defined on all of Y.

9.7 Restrictions Revisited The values of a left inverse $g : Y \to X$ of an injective $f : X \to Y$ are, however, uniquely defined on the range $f(X)$ of f. One may therefore restrict to $Z = f(X) \subset Y$ and consider instead the inverse pair $f : X \to Z$ and $g : Z \to X$. Note there is a slight abuse of notation here. The category-theoretic axiom of mutual exclusiveness of hom-sets ($\mathbf{C}(A,B) \cap \mathbf{C}(C,D) = \varnothing$ unless $A = C$ and $B = D$; cf. Definition 0.1($c1$)) implies that $f : X \to Y$ and $f : X \to Z$ are different mappings, and likewise for $g : Y \to X$ and $g : Z \to X$. So, for clarity, they should employ the different symbols for restrictions to codomain and domain, as defined in Section 1.26. The mapping pair under consideration, in this stricter notation, is then the restrictions $f|^Z : X \to Z$ and $g|_Z : Z \to X$. The two mappings satisfy

$$g|_Z \circ f|^Z = 1_X \quad \text{and} \quad f|^Z \circ g|_Z = 1_Z. \tag{4}$$

That is, $f|^Z$ and $g|_Z$ are each other's left *and* right inverses, whence mutual (*two-sided*) *inverses*. When a mapping f has a two-sided inverse, the inverse is uniquely determined, and is given the symbol f^{-1}. It is evident that the operation $(\cdot)^{-1}$ is an involution; i.e., $(f^{-1})^{-1} = f$. Relations (4) say that

(5) $$g|_Z = \left(f|^Z\right)^{-1} \quad \text{and} \quad f|^Z = \left(g|_Z\right)^{-1}.$$

With an encoding $f : X \to Y$, one may also ask whether the model containment $f(X) \subset Y$ is proper, i.e., if for every $y \in Y$ one has $f \vdash y$. A mapping $f : X \to Y$ is *surjective* when $f(X) = Y$, i.e. to each $y \in Y$ there is at least one element $x \in X$ with $f(x) = y$ (Definition 2.8). By definition, then, for any mapping $f : X \to Y$, the mapping obtained from the restriction of codomain to range, $f|^{f(X)}: X \to f(X)$, is surjective. A surjection $f : X \to Y$ (with $X \neq \varnothing$) has a right inverse. Specifically, such a right inverse 'chooses' for each $y \in Y$ an element $x \in X$ with $f(x) = y$. (The Axiom of Choice 2.22 may have to be invoked when there are potentially infinitely many choices.)

A mapping $f : X \to Y$ is *bijective* if it is both an injection and a surjection (Definition 2.15); that is, if and only if for each $y \in Y$ there exists exactly one $x \in X$ with $f(x) = y$. A bijection thus establishes a one-to-one correspondence between the respective elements of X and Y. A mapping is a bijection if and only if it has a two-sided inverse.

9.8 Sparsity of Inverse Mappings We saw above that when $f : X \to Y$ is injective, it defines an *inverse mapping* $f^{-1} : f(X) \to X$. In the looser (without 'restriction') notation, $f : X \to f(X)$ and $f^{-1} : f(X) \to X$ are bijections, and they compose to identity mappings:

(6) $$\begin{aligned} f^{-1} \circ f &= 1_X \\ f \circ f^{-1} &= 1_{f(X)} \end{aligned}$$

(but not necessarily $f \circ f^{-1} = 1_Y$). It is only in the sparse subset of injections among mappings that there is no information loss, that the action of an injective mapping f may be 'undone', its entailment 'inverted', through composition with its inverse f^{-1}, so that the domain X may be completely recovered.

A single-valued mapping concept usually has two or more interpretations for set-valued mappings. Injectivity is no exception:

9.9 Definition (*RL*: 2.7) A set-valued mapping $F : X \multimap Y$ is

i. *semi-single-valued* if, for $x_1 \neq x_2$ in X,

(7) $$F(x_1) \cap F(x_2) \neq \varnothing \quad \Rightarrow \quad F(x_1) = F(x_2),$$

and

ii. *injective* if

(8) $$x_1 \neq x_2 \quad \Rightarrow \quad F(x_1) \cap F(x_2) = \varnothing$$

(contrapositively

(9) $$F(x_1) \cap F(x_2) \neq \varnothing \quad \Rightarrow \quad x_1 = x_2).$$

For a single-valued mapping $f : X \to Y$ considered as a set-valued mapping, each value $f(x)$ is equivalently a singleton set, whence $f(x_1) \cap f(x_2) \neq \varnothing$ implies value equality $f(x_1) = f(x_2)$ [cf. (7)], so a single-valued mapping is semi-single-valued. A semi-single-valued mapping $F : X \multimap Y$ defines a *partition* of its range $F(X)$; its distinct values are pairwise disjoint subsets of Y, forming the *blocks* of the partition. It also defines a partition of its domain X: one block contains all those $x \in X$ for which $F(x) = \varnothing$ [This block is empty for a single-valued mapping.], and then the rest of X (which is cor(F), containing all those $x \in X$ for which $F(x) \neq \varnothing$) is partitioned into blocks that are in one-to-one correspondence with the blocks of $F(X)$. In other words, the semi-single-valued mapping F induces an equivalence relation 'has the same F-image as', R_F, on the domain X:

(10) $$x_1 R_F x_2 \quad \text{iff} \quad F(x_1) = F(x_2)$$

(cf. *ML*: 2.19).

It is evident that an injective set-valued mapping is semi-single-valued [(9) $\Rightarrow$ (7)]. Each block of the partition of the corange of an injective set-valued mapping is a singleton subset of X. An injective single-valued mapping (for which the corange and domain coincide) is an injective set-valued mapping [(1) $\Rightarrow$ (8), (2) $\Rightarrow$ (9)].

Semi-singled-valuedness is the 'next best thing' to injectivity. An injection retains absolute element distinguishability: there is a one-to-one correspondence between domain X and range $F(X)$. For a semi-singled-valued mapping, elements in each block of the partition of the domain have the same image and are therefore indistinct thence. There is, however, at least a one-to-one correspondence of the partitions of domain and range (or, alternatively, between the quotient sets X/R_F and $F(X)/F(R_F)$; *ML*: 2.21). A generic set-valued mapping that is not even semi-single-valued loses information on yet another level.

9.10 Definition (*RL*: 2.13) For a set-valued mapping $F : X \multimap Y$, its *inverse* is the set-valued mapping $F^{-1} : Y \multimap X$ defined by, for $y \in Y$,

$$F^{-1}(y) = \{x \in X : y \in F(x)\} = \{x \in X : (x,y) \in F\}. \tag{11}$$

Since both $F(x)$ (cf. Definition 3.1) and $F^{-1}(y)$ are both defined by the membership $(x,y) \in F$, one has

9.11 Lemma (*RL*: 2.14) *Let* $F : X \multimap Y$, $x \in X$, and $y \in Y$. Then

$$y \in F(x) \quad \textit{iff} \quad x \in F^{-1}(y). \tag{12}$$

One also has $(F^{-1})^{-1} = F$. While a single-valued mapping has an inverse if and only if it is injective, *every* set-valued mapping F has an inverse F^{-1}. The inverse of a single-valued mapping, when it exists, is bijective (with the proper restrictions accounted for), but an inverse set-valued mapping has special properties only when specially endowed:

9.12 Theorem (*RL*: 2.15) *Let* $F : X \multimap Y$, *and* $F^{-1} : Y \multimap X$ *be its inverse. Then*

i. *if F is single-valued,* F^{-1} *is injective;*
ii. *if F is injective,* F^{-1} *is single-valued;*
iii. *if F is semi-single-valued,* F^{-1} *is semi-single-valued.*

I have already discussed the two ways to define sequential composite of set-valued mappings, the sequential composition $\circ$ (Definition 3.5) and the square product $\square$ (Definition 6.1). The two compositions and their consequences have been explicated in detail in *RL*: 3.4 *et seq*. Again, neither $F^{-1} \circ F$ nor $F^{-1} \square F$ is necessarily the identity map $1_X : X \multimap X$. Thus the conventional usage of the term '*inverse* set-valued mapping' to refer to F^{-1} is a bit of a misnomer. It is more properly the 'converse'. It is not the *operational inverse*, the usual algebraic definition in connection with a 'reversal entity for the recovery of the identity'.

On the other hand, F^{-1} is the only candidate for inverse; i.e., if $G \circ F = 1_X$ (or $G \sqcap F = 1_X$), then $G = F^{-1}$.

The set-valued mappings F and F^{-1} are not guaranteed to be operational inverses even when one (hence also the other, by Theorem 9.12.iii above) is semi-singled-valued. Operational invertibility for *single-valued* set-valued mappings is, however, equivalent to injectivity:

9.13 Theorem *A single-valued mapping $F : X \multimap Y$ has an operational inverse, i.e., $F^{-1} \circ F = 1_{\mathrm{cor}(F)}$ (and $F^{-1} \sqcap F = 1_{\mathrm{cor}(F)}$), if and only if it is injective.*

PROOF F is not injective iff there are $x_1 \neq x_2$ in X for which $F(x_1) \cap F(x_2) \neq \varnothing$. Let $y \in F(x_1) \cap F(x_2)$. Then $\{x_1, x_2\} \subset F^{-1}(y)$. But $F^{-1}(y) \subset F^{-1} \circ F(x_1)$, whence $F^{-1} \circ F(x_1) \neq \{x_1\} = 1_{\mathrm{cor}(F)}(x_1)$ (and likewise, $F^{-1}(y) \subset F^{-1} \sqcap F(x_1) \neq 1_{\mathrm{cor}(F)}(x_1)$). □

Note that $F^{-1} \circ F$ and $F^{-1} \sqcap F$ are $1_{\mathrm{cor}(F)}$, and not 1_X; if $x \in \mathrm{cor}(F)$ for an injective mapping F, then $F^{-1} \circ F(x) = F^{-1} \sqcap F(x) = \{x\}$. Also, the mappings in the compositions are restricted to $\mathrm{cor}(F) \subset X$, i.e., in a more proper notation one has $F^{-1}|^{\mathrm{cor}(F)} \circ F|_{\mathrm{cor}(F)} = 1_{\mathrm{cor}(F)}$ and $F^{-1}|^{\mathrm{cor}(F)} \sqcap F|_{\mathrm{cor}(F)} = 1_{\mathrm{cor}(F)}$. Elements $x \in X$ for which $F(x) = \varnothing$ are not recoverable by F^{-1}; indeed, if $F(x) = \varnothing$, then $F^{-1} \circ F(x) = \bigcup_{y \in \varnothing} F^{-1}(y) = \varnothing$ and $F^{-1} \sqcap F(x) = \bigcap_{y \in \varnothing} F^{-1}(y) = X$. With all these caveats in place, one may conclude that an injective set-valued mapping F is a *right inverse* of F^{-1}, and that F^{-1} is a *left inverse* of F (cf. Definition 9.5). As is the case for single-valued mappings, the left–right roles of inverses for set-valued mappings are also not in general interchangeable. The property of injectivity of F is not necessarily inherited by F^{-1} (F^{-1} is single-valued by Theorem 9.12.ii, but not necessarily injective.), thus even when $F^{-1} \circ F = 1_{\mathrm{cor}(F)}$ and $F^{-1} \sqcap F = 1_{\mathrm{cor}(F)}$, one may still have $F \circ F^{-1} \neq 1_{\mathrm{cor}(F^{-1})}$ and $F \sqcap F^{-1} \neq 1_{\mathrm{cor}(F^{-1})}$.

There are many category-theoretic definitions of special morphisms that generalize injection (and surjection) in **Set**. The reader is encouraged to consult *ML*: A.41–A.46.

9.14 The Decoding Functor Commutativity of the modelling relation arrow diagram

(13)

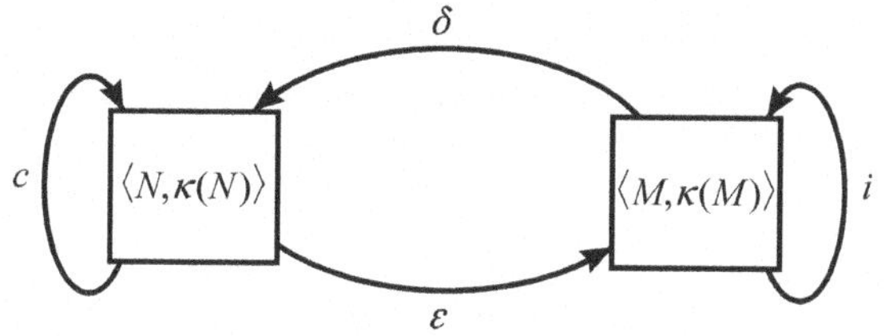

(cf. Section 5.13), whence the commutativity condition

(14) $$c = \delta \circ i \circ \varepsilon,$$

if taken in the strictest set-theoretic sense of element equality (cf. Axiom of Extension; *ML*: 0.2), would imply that the decoding functor δ has to satisfy, for each set $A \in N$ and for each mapping $F \in \mathbf{C}(A, B) \subset \kappa(N)$,

(15) $$\begin{cases} \delta \circ \varepsilon(A) = A \in \delta \circ \varepsilon(N) \subset N \subset \mathfrak{O}\mathbf{C}, \\ \delta \circ \varepsilon(F) = F \in \mathbf{C}(A, B) \subset \delta(\kappa(M)) \subset \kappa(N) \subset \mathfrak{A}\mathbf{C}. \end{cases}$$

Condition (15) is in fact the statement

(16) $$\delta \circ \varepsilon = I_{\langle N, \kappa(N) \rangle},$$

which requires the functorial composite $\delta \circ \varepsilon$ to be *equal* to the identity functor $I_{\langle N, \kappa(N) \rangle}$. Equation (16) says that the decoding functor δ is a *left inverse* of the encoding functor ε and ε is a *right inverse* of δ. Componentwise, the decoding criteria are

(17) $$\begin{cases} \delta \circ \varepsilon = 1_{\mathfrak{O}\mathbf{C}} \\ \delta \circ \varepsilon = 1_{\mathfrak{A}\mathbf{C}} \end{cases}.$$

These are very stringent requirements: the decoding functor δ has to be an operational inverse of the encoding functor ε on both objects and morphisms, whence entailing ε to be injective on objects and injective on arrows (Definition 0.8). A fortiori, with the looser notation adopted in Section 9.8, the (restricted) encoding functor $\varepsilon : \langle N, \kappa(N) \rangle \to \langle \varepsilon(N), \varepsilon(\kappa(N)) \rangle \subset \langle M, \kappa(M) \rangle$ and the (restricted) decoding functor $\delta : \langle \varepsilon(N), \varepsilon(\kappa(N)) \rangle \to \langle N, \kappa(N) \rangle$ are isomorphisms in the category **Cat**, that

$$(18) \qquad \begin{aligned} \delta \circ \varepsilon &= I_{\langle N, \kappa(N) \rangle} \\ \varepsilon \circ \delta &= I_{\langle \varepsilon(N), \varepsilon(\kappa(N)) \rangle} \end{aligned} .$$

It is, however, important to note that the chirality of the modelling relation (cf. Section 5.10) is still in place: one does *not* necessarily have $\varepsilon \circ \delta = I_{\langle M, \kappa(M) \rangle}$. Stated otherwise, the left–right inverse roles of (the unrestricted functors) δ and ε are not automatically reversible.

We saw above that the invertibility of mappings requires the exiguous injectivity, but the paucity would make the plausibility of decoding undesirably special in the world of modelling. Remedies are in order. If one were to decode more encodings with alacrity, one must relax the definition of invertibility, so as to expand the subclass of invertible mappings. This is akin to the development of *regular* summability methods in mathematical analysis, in which one redefines the notion of convergence so some divergent series then converge (while those originally convergent series remain so).

Beyond Isomorphism

9.15 Up to Isomorphism The requirement of *equality* between two objects and between two morphisms of a category is unnecessarily strict; *isomorphism* (Definition 0.5) usually suffices to serve the notion of 'sameness'.

An isomorphism in a category **C** is a morphism defined precisely by the existence of a (two-sided) inverse: A morphism $f : A \to B$ is an *isomorphism* if there exists an *inverse morphism* $g : B \to A$ such that

$$(19) \qquad \begin{aligned} f \circ g &= 1_B \\ g \circ f &= 1_A \end{aligned} .$$

If such an inverse morphism exists it is unique, and is denoted by f^{-1}. Isomorphisms in the category **Set** are bijections.

Informally, two mathematical entities of similar nature are said to be *isomorphic* if there is a one-to-one correspondence between them that preserves all relevant properties, i.e. if there is a 'structure-preserving bijection'. An isomorphism usually coincides intuitively with the most natural concept of structural preservation. The notion of 'sameness' of two objects is, in other words, as is standard in category theory, that of 'up to isomorphism'. Isomorphic objects are considered abstractly (and often identified as) the same. Most constructions of category theory are 'unique up to isomorphism' (in the sense that two similarly constructed objects are isomorphic, if not necessarily identical).

9.16 Functor as Isomorphism Recall (Definition 0.7) that a functor $F : \mathbf{C} \to \mathbf{D}$ consists of a pair of mappings $\langle F : \mathfrak{O}\mathbf{C} \to \mathfrak{O}\mathbf{D}, F : \mathfrak{A}\mathbf{C} \to \mathfrak{A}\mathbf{D} \rangle$ on the categorical components of objects and morphisms, called respectively the object mapping and the arrow mapping. In the category **Cat** of categories and (covariant) functors, an *isomorphism* $F : \mathbf{C} \to \mathbf{D}$ is a functor that is bijective both on objects and on arrows (i.e., both the object mapping $F : \mathfrak{O}\mathbf{C} \to \mathfrak{O}\mathbf{D}$ and the arrow mapping $F : \mathfrak{A}\mathbf{C} \to \mathfrak{A}\mathbf{D}$ are bijections that are **Set**-isomorphisms). When $F : \mathbf{C} \to \mathbf{D}$ is an isomorphism, the categories **C** and **D** are *isomorphic*. This is equivalent to the existence of an 'inverse functor' $G = F^{-1} : \mathbf{D} \to \mathbf{C}$, with

$$(20) \qquad \begin{array}{c} F \circ G = I_{\mathbf{D}} \\ G \circ F = I_{\mathbf{C}} \end{array}.$$

These equalities say the arrow diagram

$$(21) \qquad \mathbf{C} \underset{F}{\overset{G}{\leftrightarrows}} \mathbf{D}$$

commutes.

Objects of the functor category $\mathbf{D}^{\mathbf{C}}$ (Definition 0.19) are functors from **C** to **D**. Instead of equality, the less stringent natural isomorphism (Definition 0.20) often suffices to qualify 'sameness of functors'. Consequently, when the notion of 'up to isomorphism' is used to characterize $\mathbf{D}^{\mathbf{C}}$-objects, a more natural notion of 'sameness of categories' may be obtained. In a typically category-theoretic process of hierarchical iteration, one may, in the category **Cat**, replace, in the defining equations (20) of **Cat**-isomorphism, the strict equalities themselves by natural isomorphisms. Note that $F \circ G = I_{\mathbf{D}}$ is an equality of objects in the functor category $\mathbf{D}^{\mathbf{D}}$, and $G \circ F = I_{\mathbf{C}}$ is an equality of objects in the functor category $\mathbf{C}^{\mathbf{C}}$. This expansion to, as it were, 'isomorphism up to isomorphism', entails the notion of equivalence of categories [Mac Lane 1997, Chapter IV, Section 4]:

9.17 Definition An *equivalence* between categories **C** and **D** consists of a pair of functors $F : \mathbf{C} \to \mathbf{D}$ and $G : \mathbf{D} \to \mathbf{C}$, viz.

$$(22) \qquad \mathbf{C} \underset{F}{\overset{G}{\leftrightarrows}} \mathbf{D},$$

and a pair of natural isomorphisms

(23) $$\eta : I_{\mathbf{D}} \to F \circ G \quad \text{and} \quad \xi : G \circ F \to I_{\mathbf{C}}.$$

If there exists an equivalence between **C** and **D**, one says that **C** and **D** are *equivalent*, and writes $\mathbf{C} \simeq \mathbf{D}$. One also says that the functors F and G are *equivalences* (to each other), or even, loosely, that 'an equivalence functor' (i.e. F or G by itself) is 'an equivalence of categories'.

The condition that the two morphisms in (23) are *natural isomorphisms* is a 'shorthand'. The verbose version is

i. that there is a pair of *natural transformations* (23) (Definition 0.18), and
ii. that they are *isomorphisms* in their respective functor categories:

(24) $$\begin{array}{ll} F \circ G \cong I_{\mathbf{D}} & \left(\text{in } \mathbf{D}^{\mathbf{D}}\right) \\ G \circ F \cong I_{\mathbf{C}} & \left(\text{in } \mathbf{C}^{\mathbf{C}}\right) \end{array}.$$

Compare (24) with (20). Equalities (20) imply that the arrow diagram (21) commutes. Isomorphisms (24) do *not* imply that the arrow diagram (22) commutes (since commutativity requires *equality* of traces on paths with same initial and final vertices).

That the natural transformations (23) are natural isomorphisms implies that their components are isomorphisms in their respective categories: for each **D**-object X, $\eta_X \in \mathbf{D}(X, FGX)$ is an isomorphism in **D**, and for each **C**-object A, $\xi_A \in \mathbf{C}(GFA, A)$ is an isomorphism in **C**. Functors that are equivalences are each other's 'inverse up to isomorphism'. Because of this '(only) unique up to isomorphism' characterization, however, knowledge of F is usually not enough to reconstruct G and the natural isomorphisms η and ξ : there may be many, naturally, equivalent choices.

An alternate characterization of equivalence functors is the:

9.18 Theorem *A functor is an equivalence if and only if it is*

i. *full,*
ii. *faithful, and*
iii. *essentially surjective on objects.*

The Prolegomenon contains the definitions of the three individually necessary and jointly sufficient properties of an equivalence functor; here is a succinct summary. A functor $F : \mathbf{C} \to \mathbf{D}$ is *surjective on objects* if the object mapping $F : \mathcal{O}\mathbf{C} \to \mathcal{O}\mathbf{D}$ is surjective; i.e., for each **D**-object X there exists a **C**-object A such that $X = FA$. When the requirement of **D**-object-equality is relaxed to **D**-isomorphism, one generalizes the property: the functor $F : \mathbf{C} \to \mathbf{D}$ is *essentially* *surjective on objects* if for each **D**-object X there exists a **C**-object A such that $X \cong FA$ (Definition 0.8). For each pair of **C**-objects A and B, the functor F : $\mathbf{C} \to \mathbf{D}$ assigns to each **C**-morphism $f \in \mathbf{C}(A, B)$ a **D**-morphism $Ff \in \mathbf{D}(FA, FB)$, and so defines a (single-valued) mapping (**Set**-morphism)

$$F_{A,B} : \mathbf{C}(A, B) \to \mathbf{D}(FA, FB) \tag{25}$$

with $F_{A,B}(f) = Ff$. The functor F is *faithful* when each $F_{A,B}$ is injective, and *full* when each $F_{A,B}$ is surjective (Definition 0.14). That an equivalence functor F is faithful and full means, of course, that each $F_{A,B}$ is bijective (but this is insufficient to imply functorial isomorphism; for an explanation see Section 0.14).

9.19 Skeleton If **C** is a subcategory of **D** (Definition 0.6), then the inclusion functor (Definition 0.15) is faithful; it is full if and only if **C** is a full subcategory of D. If $F : \mathbf{C} \to \mathbf{D}$ is a faithful and full functor, then **C** is equivalent to the full subcategory $F\mathbf{C}$ of **D** (where an $F\mathbf{C}$-object has the form FA for some **C**-object A). In particular, let **C** be a full subcategory of D (whence the inclusion functor is faithful and full), and let **C** contain at least one object from each **D**-isomorphism class. Then the inclusion functor is also essentially surjective on objects, hence, by Theorem 9.18, $\mathbf{C} \simeq \mathbf{D}$. This means if one takes a category and remove from each isomorphism class some (but not all) of the objects (and their associated hom-sets), the 'slimmed-down' category is equivalent to the original. A *skeleton* is a category that is slimmed down to the point where each isomorphism class contains exactly one object.

Conversely, if one takes a category and throws in some more objects (and their associated hom-sets), each new object being isomorphic to an existing object, the one obtains a 'bulked-up' category that is also equivalent to the original. In short, categories are determined only 'up to equivalence'.

Theorem 9.18's strength allows one to demonstrate equivalence of categories without having to explicitly construct the 'inverse functor' and the associated natural isomorphisms between the composite and the identity functors (as per Definition 9.17). On the other hand, that Theorem 9.18 is an *existence* theorem is also its weakness. While the 'inverse functor' and the associated natural isomorphisms *may* be constructed (with a possible invocation of the Axiom of Choice 2.22 in the underlying set theory), the data required for the synthesis are often incomplete or multitudinous (or both).

9.20 Equivalence Entails Set-Isomorphisms Equivalence functors come in pairs. An equivalence functor $F : \mathbf{C} \to \mathbf{D}$ immanently causes a functor $G : \mathbf{D} \to \mathbf{C}$ that is an equivalence to F. If $F : \mathbf{C} \to \mathbf{D}$ is an equivalence, then Theorem 9.18 says that the hom-sets $\mathbf{C}(A, B)$ and $\mathbf{D}(FA, FB)$ are **Set**-isomorphic: for all pairs of **C**-objects A, B,

(26) $$\mathbf{C}(A,B) \cong \mathbf{D}(FA,FB),$$

and that for each **D**-object X there exists a **C**-object A such that $X \cong FA$. If $G : \mathbf{D} \to \mathbf{C}$ is an equivalence to F, then Theorem 9.18 applies likewise to G, so one also has the **Set**-isomorphisms

(27) $$\mathbf{C}(GX,GY) \cong \mathbf{D}(X,Y)$$

for all pairs of **D**-objects X, Y, and that for each **C**-object A there exists a **D**-object X such that $A \cong GX$.

9.21 Equivalence Extends Inverse If the functor $F : \mathbf{C} \to \mathbf{D}$ is an isomorphism, then it is an equivalence of categories. This is the 'regularity' requirement of the expansion of the concept of 'invertibility'. (Already-invertible functors must remain so.)

The implication

(28) $$\text{inverse} \Rightarrow \text{equivalence}$$

is because the isomorphism of F immanently entails its inverse functor $G : \mathbf{D} \to \mathbf{C}$, which together compose to $G \circ F = I_\mathbf{C}$ and $F \circ G = I_\mathbf{D}$. One may then take the natural isomorphisms $\eta : I_\mathbf{D} \to F \circ G$ and $\xi : G \circ F \to I_\mathbf{C}$ to be the *identity* natural transformations (in the functor categories $\mathbf{D}^\mathbf{D}$ and $\mathbf{C}^\mathbf{C}$ respectively).

To see that equivalence does *properly* extend (i.e., that it is not simply equal to) the concept of inverse,

(29) $$\text{Inverse} \subset \text{Equivalence},$$

one must produce an example in which $G \circ F \cong I_\mathbf{C}$ but $G \circ F \neq I_\mathbf{C}$.

Let **C** be the category of finite-dimensional vector spaces and linear transformations over the real field $\mathbb{R}$, and let **D** be the category of real matrices, with objects $\mathcal{O}\mathbf{D} = \{\mathbb{R}^n : n = 0, 1, 2, \ldots\}$ (where $\mathbb{R}^0 = \varnothing$) and morphisms $\mathbf{D}(\mathbb{R}^n, \mathbb{R}^m) =$ all $m \times n$ matrices over $\mathbb{R}$. The convention is that for any m, $\mathbf{D}(\varnothing, \mathbb{R}^m) = \{\varnothing\}$, containing exactly one **D**-morphism, the 'empty matrix', and for $n > 1$, $\mathbf{D}(\mathbb{R}^n, \varnothing) = \varnothing$ (cf. Section 2.4).

That every finite-dimensional real vector space of dimension n is isomorphic (in the sense of linear algebra) to $\mathbb{R}^n$ is a fundamental theorem in linear algebra, so is the fact that there is a one-to-one correspondence between the set of all linear transformations from an n-dimensional vector space to an m-dimensional vector space (with chosen ordered bases) and the set of all $m \times n$ matrices over $\mathbb{R}$. Let the functor $F : \mathbf{C} \rightarrow \mathbf{D}$ map an n-dimensional vector space V to $\mathbb{R}^n \in \mathfrak{O}\mathbf{D}$ and a linear transformation to its corresponding $m \times n$ matrix in $\mathbf{D}(\mathbb{R}^n, \mathbb{R}^m)$. The functor $G : \mathbf{D} \rightarrow \mathbf{C}$ maps $\mathbb{R}^n \in \mathfrak{O}\mathbf{D}$ to the vector space $\mathbb{R}^n \in \mathfrak{O}C$, and a matrix in $\mathfrak{A}\mathbf{D}$ to its corresponding linear transformation. Both functors are faithful, full, and essentially surjective on objects. But $(G \circ F)(V)$ only recovers the vector space $\mathbb{R}^n$ to which V is isomorphic, and not V itself.

9.22 Encoding and Decoding as Equivalences When one eases the equality requirement in (16) of the invertibility between the encoding and decoding functors ε and δ, $\delta \circ \varepsilon = I_{\langle N,\kappa(N)\rangle}$, to the relationship of their being equivalences, one obtains the natural isomorphism

(30) $$\delta \circ \varepsilon \cong I_{\langle N,\kappa(N)\rangle}.$$

On the one hand, this is less stringent than before. But on the other hand, functorial equivalence is a symmetric relation, so with (28) one also has

(31) $$\varepsilon \circ \delta \cong I_{\langle M,\kappa(M)\rangle},$$

whereas previously the chirality of left–right invertibility does not necessitate $\varepsilon \circ \delta = I_{\langle M,\kappa(M)\rangle}$. Since (30) and (31) are isomorphisms and not equalities, the arrow diagram of the functorial connections

(32) $$\langle N, \kappa(N)\rangle \underset{\varepsilon}{\overset{\delta}{\rightleftarrows}} \langle M, \kappa(M)\rangle$$

is not commutative (as explained above in Section 9.17), but only serves to notate the category-theoretic entities involved.

In summary, this chapter has illustrated the opposing tensions in action. The expansion of functorial invertibility by moving from equality to equivalence is a *relaxation* of requirements, but it pays a price of *restriction* that is symmetry. The modelling relation is inherently asymmetric. The next step on the invertibility journey is a means of symmetry-breaking.

10
Adjunction

Learning is but an adjunct to ourself
And where we are our learning likewise is.

— William Shakespeare (*c.* 1595)
Love's Labour's Lost
Act IV, scene iii

Asymmetry

10.1 Equivalence Symmetry Recall (Section 0.14) that, given a functor $F: \mathbf{C} \to \mathbf{D}$, to each pair of **C**-objects A and B there corresponds a (single-valued) mapping

(1) $$F_{A,B} : \mathbf{C}(A,B) \to \mathbf{D}(FA, FB)$$

defined by $F_{A,B}(f) = Ff$. The collection

(2) $$F = \{F_{A,B} : A, B \in \mathfrak{O}\mathbf{C}\}$$

constitutes the arrow mapping $F : \mathfrak{A}\mathbf{C} \to \mathfrak{A}\mathbf{D}$ of the functor (Definition 0.7) that assigns to a **C**-morphism $f \in \mathbf{C}(A,B)$ its value the **D**-morphism $Ff \in \mathbf{D}(FA, FB)$.

When the pair of functors $F : \mathbf{C} \to \mathbf{D}$ and $G : \mathbf{D} \to \mathbf{C}$ are equivalences (Definition 9.17), the mappings $F_{A,B} : \mathbf{C}(A,B) \to \mathbf{D}(FA, FB)$ (and symmetrically for all $X, Y \in \mathfrak{O}\mathbf{D}$ the mappings $G_{X,Y} : \mathbf{D}(X,Y) \to \mathbf{C}(GX, GY)$) are **Set**-isomorphisms (i.e., bijections; cf. Theorem 9.18 and Remarks 9.20). Thus one has, in **Set**,

A. H. Louie, *Intangible Life*, Anticipation Science 2,
DOI 10.1007/978-3-319-65409-6_11

(3)
$$\mathbf{C}(A,B) \cong \mathbf{D}(FA,FB)$$
$$\mathbf{C}(GX,GY) \cong \mathbf{D}(X,Y).$$

10.2 Hom-Set Definition (*ML*: A.48) An *adjunction* from category **D** to category **C** consists of a pair of functors $G : \mathbf{D} \to \mathbf{C}$ and $F : \mathbf{C} \to \mathbf{D}$, viz.

(4)
$$\mathbf{C} \underset{F}{\overset{G}{\rightleftarrows}} \mathbf{D},$$

and a family

(5)
$$\varphi = \{\varphi_{A,X} : A \in \mathbf{OC}, X \in \mathbf{OD}\}$$

of mappings

(6)
$$\varphi_{A,X} : \mathbf{C}(GX,A) \to \mathbf{D}(X,FA)$$

that are bijections (**Set**-isomorphisms)

(7)
$$\mathbf{C}(GX,A) \cong \mathbf{D}(X,FA),$$

natural in $A \in \mathbf{OC}$ and $X \in \mathbf{OD}$. The functor G is called a *left adjoint* for (or 'of') the functor F, and F is called a *right adjoint* for G (cf. the final sections in the Appendix of *ML* on adjoints, A.47–A.54). Note that, as is the case for equivalences, arrow diagram (4) simply serves to indicate the domains and codomains of the functors, and is not required to be commutative.

The left–right adjointness relationship between the two functors is denoted

(8)
$$G \dashv F.$$

Contrast the adjunction symbol $\dashv$ with the entailment symbol $\vdash$ (cf. Notations 2.2).

Compare equivalence with adjunction: Definition 9.17 with Definition 10.2. Contrast the forms of (1) with (6), and the pair of bijections (3) with the single bijection (7).

10.3 Inherent Asymmetry Equivalence functors and adjoint functors both come in pairs. Equivalence is symmetric; adjunction is asymmetric. Equivalence has a two-sided 'parity', but adjunction has a left–right 'chirality', hence the roles of left and right adjoint functors, just as those of their left and right inverse-mapping counterparts, are not generally interchangeable.

G is left adjoint to F if and only if F is right adjoint to G. The asymmetry is inherent in the definition of adjunction itself, that it is *from* category **D** *to* category **C**, the flow direction of the left adjoint $G : \mathbf{D} \to \mathbf{C}$. The causes of asymmetry in adjointness may be further explicated thus. There is a directionality in the hom-sets that appear in the adjunction isomorphisms (7). Morphisms in $\mathbf{C}(GX, A)$ and $\mathbf{D}(X, FA)$ are respectively $GX \to A$ and $X \to FA$, with *material* causes $GX \in G\mathbf{D}$ and $X \in \mathbf{D}$, and *final* causes $A \in \mathbf{C}$ and $FA \in F\mathbf{C}$. The *efficient* cause of the morphisms is hence the flow from **D** to **C**, that of the *formal* cause of the left adjoint $G : \mathbf{D} \to \mathbf{C}$. The four Aristotelian causes therefore concertedly make the left adjoint $G : \mathbf{D} \to \mathbf{C}$ 'that which is entailed' and the right adjoint $F : \mathbf{C} \to \mathbf{D}$ 'that which entails'. The entailment allusion of adjunction implied by the notations, that 'right adjoint $\vdash$ left adjoint' in (8), is whence not coincidental. Functors have left adjoints more often than right adjoints. If the congenial information loss of mappings is the manifestation in mathematics of thermodynamic entropy in nature, then this inherent left–right asymmetry in adjunction is the manifestation in mathematics of parity violation in nature.

Unit and Counit

Adjunction may equivalently be defined thus:

10.4 Counit–Unit Definition An *adjunction* [Mac Lane 1997, Chapter IV, Section 1] from category **D** to category **C** consists of a pair of functors $G : \mathbf{D} \to \mathbf{C}$ and $F : \mathbf{C} \to \mathbf{D}$, and a pair of natural transformations

(9) $$\eta : I_{\mathbf{D}} \to F \circ G \quad \text{and} \quad \bar{\eta} : G \circ F \to I_{\mathbf{C}},$$

respectively called the *unit* and the *counit* of the adjunction, such that for each $A \in \mathcal{O}\mathbf{C}$,

(10) $$F\bar{\eta}_A \circ \eta_{FA} = 1_{FA} \in \mathbf{D}(FA, FA),$$

and for each $X \in \mathcal{O}\mathbf{D}$,

(11) $$\bar{\eta}_{GX} \circ G\eta_X = 1_{GX} \in \mathbf{C}(GX, GX).$$

The compositions (10) and (11) may be represented as commuting triangles (of morphisms respectively in **D** and **C**):

(12) 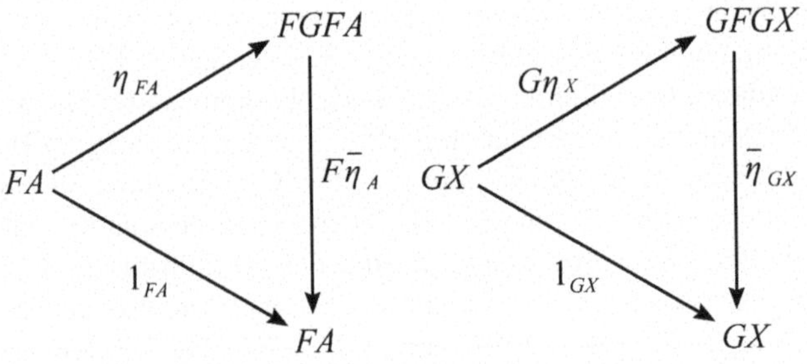

The unit $\eta : I_{\mathbf{D}} \to F \circ G$ is a natural transformation (a morphism in the category $\mathbf{D}^{\mathbf{D}}$, $\eta \in \mathbf{D}^{\mathbf{D}}(I_{\mathbf{D}}, F \circ G)$). The component η_X of η at $X \in \mathcal{O}\mathbf{D}$ is the $\mathbf{D}$-morphism $\eta_X \in \mathbf{D}(I_{\mathbf{D}}X, (F \circ G)X) = \mathbf{D}(X, FGX)$, which may be defined through the adjunction bijection (7), $\varphi_{GX,X} : \mathbf{C}(GX, GX) \to \mathbf{D}(X, FGX)$, evaluated at 1_{GX}:

$$\eta_X = \varphi_{GX,X}(1_{GX}). \tag{13}$$

The commutativity (Property ($t1$) of Definition 0.18) of natural-transformation-defined morphisms, in the context of the unit $\eta : I_{\mathbf{D}} \to F \circ G$, for $g \in \mathbf{D}(X, Y)$, is $FG(g) \circ \eta_X = \eta_Y \circ g$:

(14) 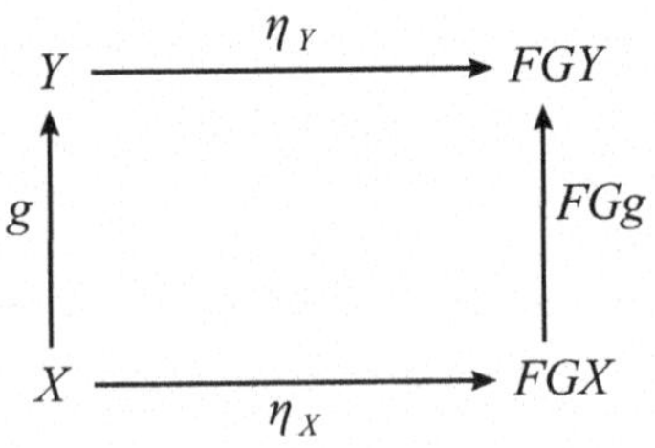

The counit is conventionally given the symbol ε, but since I am using ε for the encoding functor, I denote the counit by the symbol $\bar{\eta}$ that makes the connection of the unit–counit pair $\langle \eta, \bar{\eta} \rangle$ even more apparent. The counit $\bar{\eta} : G \circ F \to I_{\mathbf{C}}$ is a natural transformation (a morphism in the category $\mathbf{C}^{\mathbf{C}}$, $\bar{\eta} \in \mathbf{C}^{\mathbf{C}}(G \circ F, I_{\mathbf{C}})$). The component $\bar{\eta}_A$ of $\bar{\eta}$ at $A \in \mathcal{O}\mathbf{C}$ is the $\mathbf{C}$-morphism

$\bar{\eta}_A \in \mathbf{C}((G \circ F)A, I_{\mathbf{C}}A) = \mathbf{C}(GFA, A)$, which may be defined through the *inverse* adjunction bijection $\varphi^{-1}_{A,FA} : \mathbf{D}(FA, FA) \to \mathbf{C}(GFA, A)$ evaluated at 1_{FA}:

(15)
$$\bar{\eta}_A = \varphi^{-1}_{A,FA}(1_{FA}).$$

Dually, the commutativity of natural-transformation-defined morphisms, when specialized to the counit $\bar{\eta} : G \circ F \to I_{\mathbf{C}}$, for $f \in \mathbf{C}(A, B)$, is $\bar{\eta}_B \circ GF(f) = f \circ \bar{\eta}_A$:

(16)

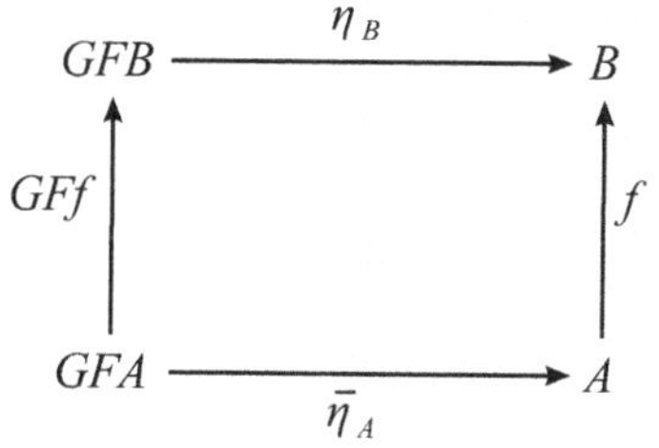

10.5 The Counit–Unit Equations The relation (10) of morphisms is natural in A, the relation (11) is natural in X, and the morphisms involved are components of natural transformations. Accordingly, the relations may be abbreviated by removing their respective variables, and become the following equations of natural transformations:

(17)
$$\begin{array}{ll} F\bar{\eta} \circ \eta_F = 1_F & (\text{in } \mathbf{D}^{\mathbf{C}}) \\ \bar{\eta}_G \circ G\eta = 1_G & (\text{in } \mathbf{C}^{\mathbf{D}}) \end{array}.$$

These are called the *counit–unit equations*. In natural transformation form, the commuting triangles (12) become

(18)

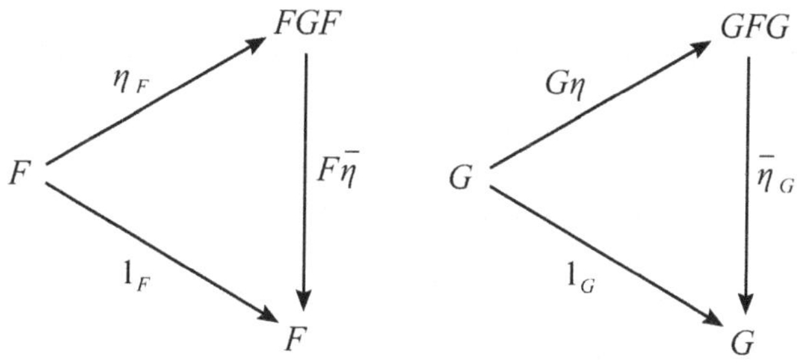

In view of the geometry of these arrow diagrams of the compositions, equations (17) are sometimes called 'triangular identities' [Mac Lane 1997, Chapter IV, Section 1]. The asymmetry in the counit–unit equations (and of the natural transformations (9) themselves) imparts a chirality on the two functors, and accordingly the functor G is called a *left adjoint* for the functor F, and F is called a *right adjoint* for G. An adjunction may then be denoted $\langle G, F, \eta, \bar{\eta} \rangle$, an ordered quadruple of ⟨left adjoint, right adjoint, unit, counit⟩.

10.6 Adjunction Versus Equivalence Compare Definition 10.4 with Definition 9.17; note that the requisite pairs of natural transformations, that for adjunction in (9) and that for equivalence,

$$\eta : I_{\mathbf{D}} \to F \circ G \quad \text{and} \quad \xi : G \circ F \to I_{\mathbf{C}}, \tag{19}$$

have the same form. (The reason for using different symbols, $\langle \eta, \bar{\eta} \rangle$ versus $\langle \eta, \xi \rangle$, will be explained presently.) Also contrast the counit–unit equations (17) with the requisite isomorphisms for equivalence

$$\begin{aligned} F \circ G &\cong I_{\mathbf{D}} \quad (\text{in } \mathbf{D}^{\mathbf{D}}) \\ G \circ F &\cong I_{\mathbf{C}} \quad (\text{in } \mathbf{C}^{\mathbf{C}}) \end{aligned} . \tag{20}$$

Equivalence is a relation between categories, and (20) is a statement on when functors compose to identity functors. Adjunction is a relation between functors, and (17) is a statement on when natural transformations compose to identity natural transformations.

Beyond Equivalence

10.7 Adjunction Generalizes Equivalence One may consider adjunction the 'next step' after equivalence in the relaxation of the concept of 'functorial inverse': an equivalence functor is an adjoint.

Let the functors $F : \mathbf{C} \to \mathbf{D}$ and $G : \mathbf{D} \to \mathbf{C}$, with the pair of natural isomorphisms $\eta : I_{\mathbf{D}} \to F \circ G$ and $\xi : G \circ F \to I_{\mathbf{C}}$, be an equivalence of categories. For each $X \in \mathbf{OD}$, the component $\eta_X \in \mathbf{D}(X, FGX)$ is an isomorphism, whence for $X, Y \in \mathbf{OD}$, $\mathbf{D}(X,Y) = \mathbf{D}(FGX, Y) \times \mathbf{D}(X, FGX) \cong \mathbf{D}(FGX, Y)$. Thus for $A \in \mathbf{OC}$ and $X \in \mathbf{OD}$,

$$\mathbf{C}(GX, A) \xrightarrow{F} \mathbf{D}(FGX, FA) \longleftrightarrow \mathbf{D}(X, FA); \tag{21}$$

i.e., $\mathbf{C}(GX, A) \cong \mathbf{D}(X, FA)$. This entails the adjunction $G \dashv F$ by Definition 10.2. Dually, for each $A \in \mathbf{OC}$, the component $\xi_A \in \mathbf{C}(GFA, A)$ is an isomorphism, whence for $A, B \in \mathbf{OC}$, $\mathbf{C}(GFA, B) = \mathbf{C}(A, B) \times \mathbf{C}(GFA, A) \cong \mathbf{C}(A, B)$. Thus for $A \in \mathbf{OC}$ and $X \in \mathbf{OD}$,

$$\mathbf{D}(FA, X) \xrightarrow{G} \mathbf{C}(GFA, GX) \longleftrightarrow \mathbf{C}(A, GX); \tag{22}$$

i.e., $\mathbf{D}(FA, X) \cong \mathbf{C}(A, GX)$. This entails the adjunction $F \dashv G$ by Definition 10.2.

In sum, each equivalence functor in the pair $\langle G, F \rangle$ is both the left and the right adjoint of the other.

10.8 Adjoint Equivalence One must note, however, that because of the 'up-to-isomorphism' characterization, it is not necessary that the original pair $\langle \eta, \xi \rangle$ of natural isomorphisms satisfy the counit–unit equations (17), which are *equalities* of natural transformations [and equalities of morphisms in component form (10) and (11)]. This explains why I used $\langle \eta, \xi \rangle$ in (19) (and in Definition 9.17), but not $\langle \eta, \bar{\eta} \rangle$.

Given equivalence functors $F : \mathbf{C} \to \mathbf{D}$ and $G : \mathbf{D} \to \mathbf{C}$ and a natural isomorphism $\eta : I_{\mathbf{D}} \to F \circ G$, however, one may *construct* a unique natural isomorphism $\bar{\eta} : G \circ F \to I_{\mathbf{C}}$ so that $\langle G, F, \eta, \bar{\eta} \rangle$ would satisfy the counit–unit equations (dually, from a natural isomorphism $\xi : G \circ F \to I_{\mathbf{C}}$ one may construct its corresponding unit $\underline{\xi} : I_{\mathbf{D}} \to F \circ G$). This construction yields an adjunction that is simultaneously an equivalence, and is called an *adjoint equivalence of categories* [Mac Lane 1997, Chapter IV, Section 4].

In an adjoint equivalence $\langle G, F, \eta, \bar{\eta} \rangle$, the functors $F : \mathbf{C} \to \mathbf{D}$ and $G : \mathbf{D} \to \mathbf{C}$ are 'two-sided adjoints' of each other. When $\eta : I_{\mathbf{D}} \to F \circ G$ and $\bar{\eta} : G \circ F \to I_{\mathbf{C}}$ are natural isomorphisms, then so are their inverses $\eta^{-1} : F \circ G \to I_{\mathbf{D}}$ and $\bar{\eta}^{-1} : I_{\mathbf{C}} \to G \circ F$. With the latter pair as counit and unit, the counit–unit equations (17) may then be written as

(23)
$$\begin{aligned} G\eta^{-1} \circ \bar{\eta}_G^{-1} &= 1_G \quad (\text{in } \mathbf{C}^{\mathbf{D}}) \\ \eta_{\bar{F}}^{-1} \circ F\bar{\eta}^{-1} &= 1_F \quad (\text{in } \mathbf{D}^{\mathbf{C}}) \end{aligned},$$

and then say that $\langle \bar{\eta}^{-1}, \eta^{-1} \rangle$ is the unit–counit pair for the adjunction between the left functor $F : \mathbf{C} \to \mathbf{D}$ and right functor $G : \mathbf{D} \to \mathbf{C}$. Thus $\langle F, G, \bar{\eta}^{-1}, \eta^{-1} \rangle$ is the *dual adjoint equivalence of categories.*

10.9 Abelian and Tauberian Theorems We see, therefore, that adjunction is a furtherance in the expansive sequence of equality–isomorphism–equivalence towards a more generalized notion of invertibility. The expansions are 'regular', in the sense (borrowed from summability theory) that each new expansion concept encompasses the old. Again borrowing the terminology from summability theory, we may call results of such inclusive expansion 'Abelian theorems'.

'Partial converse results' are called 'Tauberian theorems', when a new concept together with a side-condition imply an old method. (If an inclusive expansion by itself implies the old concept, then it is simply identical to the old and thence no expansion at all.) Here is an example—if either the left or right adjoint functor in an adjunction is also faithful and full, then so is the other in the pair, and they are equivalence functors (hence part of an adjoint equivalence).

10.10 Theorem *The following statements are equivalent for functors* $F : \mathbf{C} \to \mathbf{D}$ *and* $G : \mathbf{D} \to \mathbf{C}$:

i. *G is a left adjoint of F and both functors are faithful and full.*
ii. *F is a right adjoint of G and both functors are faithful and full.*
iii. *There are natural isomorphisms*

(24)
$$\eta : I_{\mathbf{D}} \to F \circ G \quad \text{and} \quad \bar{\eta} : G \circ F \to I_{\mathbf{C}}.$$

Compare Theorem 10.10 with Definition 9.17 and Theorem 9.18.

10.11 Modelling as Adjunction In the context of the modelling relation

(25)
$$\langle N, \kappa(N) \rangle \underset{\varepsilon}{\overset{\delta}{\leftrightarrows}} \langle M, \kappa(M) \rangle,$$

adjunction is

(26) $$\delta \dashv \varepsilon.$$

The encoding functor $\varepsilon : \langle N, \kappa(N)\rangle \to \langle M, \kappa(M)\rangle$ is the right adjoint in search of a decoding functor $\delta : \langle M, \kappa(M)\rangle \to \langle N, \kappa(N)\rangle$ as its the left adjoint. The natural isomorphism

(27) $$\delta \circ \varepsilon \cong I_{\langle N, \kappa(N)\rangle}$$

of equivalence is relaxed to the natural transformation

(28) $$\bar{\eta} : \delta \circ \varepsilon \to I_{\langle N, \kappa(N)\rangle}$$

that is the counit of the adjunction.

The expansion of functorial invertibility by moving from equivalence to adjunction restores the inherently asymmetry of the modelling relation. The 'right adjoint $\vdash$ left adjoint' entailment provides a way to construct the left adjoint functor from a given right adjoint functor. (In the next few chapters I shall present examples.) A model is only as useful as the amount of information that can be decoded, and the '$\varepsilon \vdash \delta$.' entailment is the means to explicitly construct decodings.

Adjointness as a Universal Property

10.12 Universality In various branches of mathematics, many constructions are characterized by properties that require an object X to be 'special' among all other similar objects, in the sense that if another object Y satisfies similar conditions, then there exists a unique structure-preserving mapping between that other object Y and the special object X. These properties are called *universal properties* (*ML*: A.20, A.31) and optimize the quality of 'generality'. The existence of the structure-preserving mapping says that "X is general enough", while the uniqueness of the structure-preserving mapping says that "X is not too general". A universal property is *initial* if the mapping involved initiates from the special object X (hence initial object and initial morphism; *ML*: A.35). A universal property is *terminal* if the mapping involved terminates on the special object X (thence terminal object and terminal morphism; *ML*: A.25).

In the language of category theory, a construction is *most efficient* if it satisfies a universal property, and is *formulaic* if it defines a functor. Adjunction may equivalently be formulated via universal properties. The search for and construction of an adjoint functor are then intuitively analogous to an optimization procedure.

10.13 Universal Property Definition (*ML*: A.49) $F : \mathbf{C} \to \mathbf{D}$ is a *right adjoint functor* (for $G : \mathbf{D} \to \mathbf{C}$) if it satisfies the universal property that for each D-object X, there exists an *initial morphism* from X to F. Explicitly, there is a C-object GX and a D-morphism $\eta_X : X \to FGX$, such that for any D-morphism $g : X \to FA$, there is a unique C-morphism $f : GX \to A$ satisfying $Ff \circ \eta_X = g$:

(29)

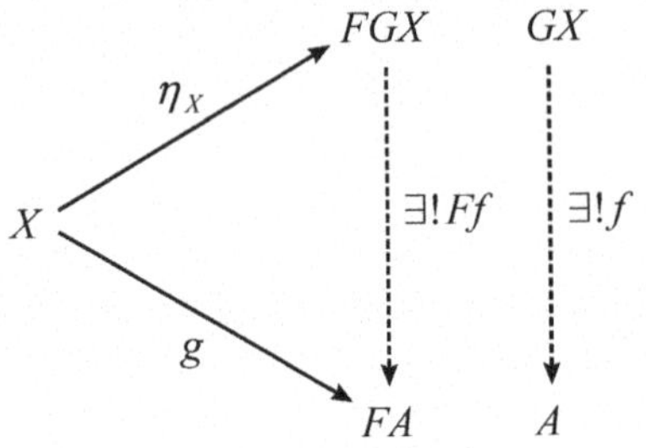

Dually, $G : \mathbf{D} \to \mathbf{C}$ is a *left adjoint functor* (for $F : \mathbf{C} \to \mathbf{D}$) if it satisfies the universal property that for each **C**-object A, there exists a *terminal morphism* from G to A. There is a **D**-object FA and a **C**-morphism $\xi_A : GFA \to A$, such that for any **C**-morphism $f : GX \to A$, there is a unique **D**-morphism $g : X \to FA$ satisfying $\xi_A \circ Gg = f$:

(30)

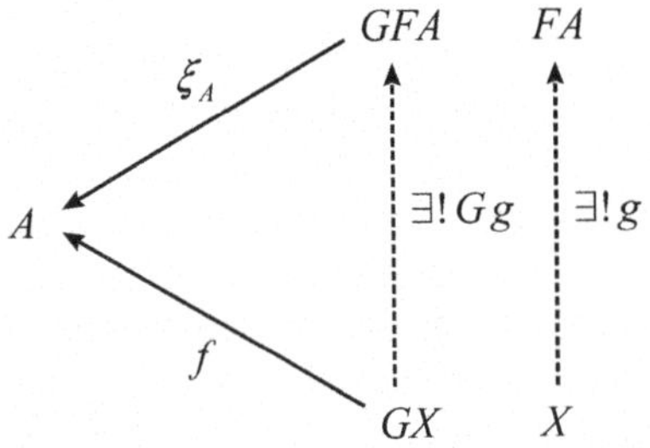

10.14 Adjunction Components From the three equivalent Definitions 10.2, 10.4, and 10.13 of adjunction, one sees that there are many category-theoretic entities associated with the adjunction $G \dashv F$.

To wit, an *adjunction* from category $\mathbf{D}$ to category $\mathbf{C}$ consists of

i. a functor $G : \mathbf{D} \to \mathbf{C}$ called the *left adjoint*;
ii. a functor $F : \mathbf{C} \to \mathbf{D}$ called the *right adjoint*;
iii. a family $\varphi = \{\varphi_{A,X} : A \in \mathbf{OC}, X \in \mathbf{OD}\}$ of bijective mappings (i.e., **Set**-*isomorphisms*) $\varphi_{A,X} : \mathbf{C}(GX, A) \to \mathbf{D}(X, FA)$;
iv. a natural transformation $\eta : I_{\mathbf{D}} \to F \circ G$ called the *unit*, with component $\eta_X \in \mathbf{D}(X, FGX)$ at $X \in \mathbf{OD}$.
v. a natural transformation $\bar{\eta} : G \circ F \to I_{\mathbf{C}}$ called the *counit*, with component $\bar{\eta}_A \in \mathbf{C}(GFA, A)$ at $A \in \mathbf{OC}$;

These entities are related thus: with $A \in \mathbf{OC}$ and $X \in \mathbf{OD}$, for every $\mathbf{C}$-morphism $f : GX \to A$ there is a unique $\mathbf{D}$-morphism $\varphi_{A,X}(f) = g : X \to FA$, and for every $\mathbf{D}$-morphism $g : X \to FA$ there is a unique $\mathbf{C}$-morphism $\varphi_{A,X}^{-1}(g) = f : GX \to A$, such that the following diagrams commute:

(31)

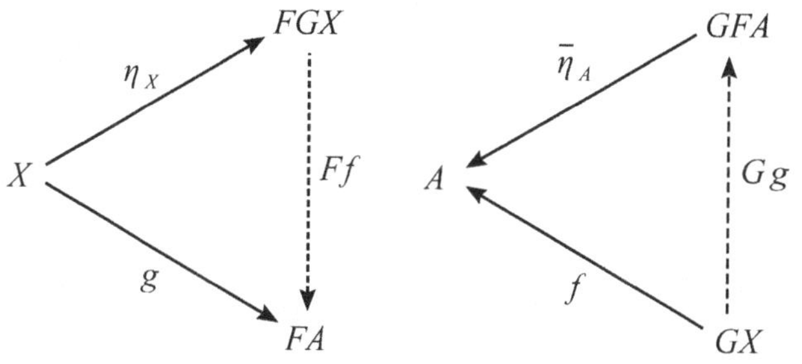

Explicitly, their formulaic relations are

(32)
$$\begin{aligned} g &= \varphi_{A,X}(f) = Ff \circ \eta_X && \in \mathbf{D}(X, FA) \\ f &= \varphi_{A,X}^{-1}(g) = \bar{\eta}_A \circ Gg && \in \mathbf{C}(GX, A) \end{aligned}.$$

One may also recall from (13) and (15) that

(33)
$$\begin{aligned} \eta_X &= \varphi_{GX,X}(1_{GX}) && \in \mathbf{D}(X, FGX) \\ \bar{\eta}_A &= \varphi_{A,FA}^{-1}(1_{FA}) && \in \mathbf{C}(GFA, A) \end{aligned},$$

and that the pair $\langle \eta, \bar{\eta} \rangle$ of natural isomorphisms satisfy the counit–unit equations (17), that

(34)
$$\begin{aligned} F\bar{\eta} \circ \eta_F &= 1_F \quad (\text{in } \mathbf{D}^{\mathbf{C}}) \\ \bar{\eta}_G \circ G\eta &= 1_G \quad (\text{in } \mathbf{C}^{\mathbf{D}}) \end{aligned} .$$

Further, each pair $(FA, \bar{\eta}_A)$ is a terminal morphism from G to A in $\mathbf{C}$, and each pair (GX, η_X) is an initial morphism from X to F in $\mathbf{D}$.

10.15 Optimization The search for an adjoint functor is, to recapitulate, a process of optimization. Starting with a right adjoint functor $F : \mathbf{C} \to \mathbf{D}$, the construction of its left adjoint functor $G : \mathbf{D} \to \mathbf{C}$ follows the most efficient and formulaic solution given by the above relations (32)–(34).

In the modelling relation, the right adjoint F is the encoding functor $\varepsilon : \langle N, \kappa(N) \rangle \to \langle M, \kappa(M) \rangle$, and this *most efficient* and *formulaic* optimization procedure yields its left adjoint G the decoding functor $\delta : \langle M, \kappa(M) \rangle \to \langle N, \kappa(N) \rangle$.

(35)

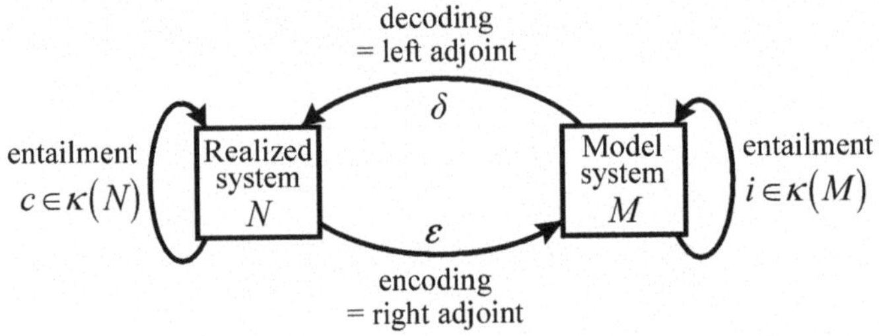

11 Descartes and Galois

The slogan is "Adjoint functors arise everywhere."

— Saunders Mac Lane (1997)
Categories for the Working Mathematician
Preface to the First Edition

We shall discover how true Mac Lane's slogan is. The remainder of *IL* contains a plethora of examples of $G \dashv F$,

Product and Exponential

Apud me omnia fiunt Mathematicè in Natura.

[In my opinion, all things in nature occur mathematically.]

— René Descartes (1640)
Letter to Marin Mersenne, 11 March 1640

On our relational biology journey, we have already encountered (cf. Section 8.9 et seq. and the references therein to *ML* and *RL*) an important adjunction, from **Set** to **Set**:

$$\mathbf{Set} \underset{F}{\overset{G}{\rightleftarrows}} \mathbf{Set}. \tag{1}$$

A. H. Louie, *Intangible Life*, Anticipation Science 2,
DOI 10.1007/978-3-319-65409-6_12

11.1 Cartesian Product Functor Fix a set S. The left adjoint functor is

(2) $$G : \begin{cases} X \mapsto X \times S & (X \in \mathfrak{O}\,\mathbf{Set}) \\ [g : X \to Y] \mapsto [Gg : X \times S \to Y \times S] & (g \in \mathfrak{A}\,\mathbf{Set}) \end{cases}.$$

It sends a set X to $X \times S$, its *Cartesian product* (Definition 1.15) with the fixed set S. Its arrow mapping is defined, for $g \in \mathbf{Set}(X, Y)$, $x \in X$, and $s \in S$, by

(3) $$(Gg)(x, s) = (g(x), s) \in Y \times S;$$

i.e., $Gg = (g,\, 1_S) = g \times 1_S$.

11.2 Hom-Functor The right adjoint functor is

(4) $$F : \begin{cases} A \mapsto A^S & (A \in \mathfrak{O}\,\mathbf{Set}) \\ [f : A \to B] \mapsto [Ff : A^S \to B^S] & (f \in \mathfrak{A}\,\mathbf{Set}) \end{cases}.$$

It assigns to each set A the set $FA = A^S = \mathbf{Set}(S, A)$, and to a mapping $f : A \to B$ the mapping $Ff : \mathbf{Set}(S, A) \to \mathbf{Set}(S, B)$ defined, for $k \in A^S = \mathbf{Set}(S, A)$, by the sequential composite

(5) $$(Ff)(k) = f \circ k \in B^S = \mathbf{Set}(S, B).$$

One sees, therefore, that F is precisely the *covariant hom-functor* $h^S = \mathbf{Set}(S, \bullet)$ from **Set** to **Set** (Definition 0.12).

The Cartesian product functor $G = \bullet \times S$ and the hom-functor $F = h^S$ are connected by the isomorphisms

(6) $$\varphi_{A,X} : \mathbf{Set}(X \times S, A) \to \mathbf{Set}(X, A^S),$$

natural in A and X (and, incidentally, also natural in S).

11.3 Bijections A mapping $u : X \times S \to A$ may be dissected to act on its two variables sequentially. The material entailment

(7) $$u \vdash a = u(x, s)$$

unfolds into a mapping $\varphi_{A,X}(u) : X \to A^S$ of the first variable $(x \in X)$ with values that are themselves mappings $(\varphi_{A,X}(u))(x) : S \to A$, followed by the application of $(\varphi_{A,X}(u))(x)$ on the second variable $(s \in S)$, yielding the final cause in codomain A:

(8) $$\{(\varphi_{A,X}(u))(x)\}(s) = u(x,s) = a \quad \text{for } x \in X \quad \text{and} \quad s \in S.$$

The unfolding of (7) is the hierarchical composition (of a functional entailment followed by a material entailment)

(9) $$\varphi_{A,X}(u) \vdash (\varphi_{A,X}(u))(x) \vdash a,$$

which has the relational diagram

(10)

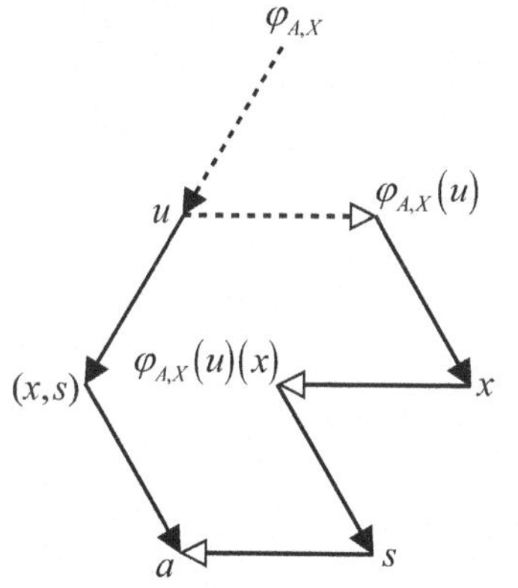

The elemental correspondence of (6),

(11) $$\varphi_{A,X} : u(\bullet,\cdot) \mapsto \{(\varphi_{A,X}(u))(\bullet)\}(\cdot),$$

is shown with dashed arrows in diagram (10).

Conversely, a functional entailment $v : X \to A^S$ may be evaluated at the variable $s \in S$ of its entailed mapping $v(x)$:

(12) $$(v(x))(s) = \left(\varphi_{A,X}^{-1}(v)\right)(x,\ s) \in A \quad \text{for } x \in X \quad \text{and} \quad s \in S.$$

The inverse correspondence

(13) $$\varphi_{A,X}^{-1} : (v(\bullet))(\cdot) \mapsto \left(\varphi_{A,X}^{-1}(v)\right)(\bullet,\cdot)$$

transforms the evaluation of the functional entailment $v : X \to A^S$ into a material entailment $\varphi_{A,X}^{-1}(v) : X \times S \to A$ with a two-variable domain:

(14)

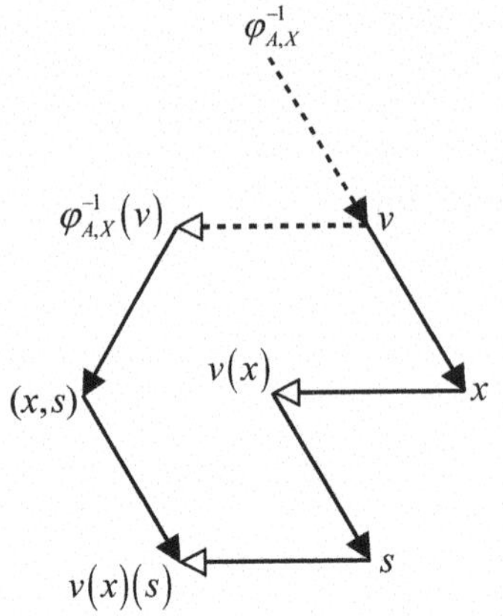

In the notation of Section 8.9, $\varphi_{A,X}(u)$ is $u^{\rightarrow}$, and $\varphi_{A,X}^{-1}(v)$ is $v^{\leftarrow}$.

11.4 Unit and Initial Morphism The component of the unit of this adjunction at set X is

(15) $$\eta_X = \varphi_{X\times S,X}(1_{X\times S}) \in \mathbf{Set}\left(X, (X \times S)^S\right),$$

the *order-pairing* defined, for $x \in X$ and $s \in S$, by

(16) $$\{\eta_X(x)\}(s) = \left\{\left(\varphi_{X\times S,X}(1_{X\times S})\right)(x)\right\}(s) = (x,s).$$

The Cartesian product set $GX = X \times S$ along with the unit component mapping $\eta_X : X \to FGX = (X \times S)^S$ are such that, for any mapping $v : X \to A^S$, there is a unique mapping $\varphi_{A,X}^{-1}(v) : X \times S \to A$ satisfying $F\varphi_{A,X}^{-1}(v) \circ \eta_X = v$:

(17)

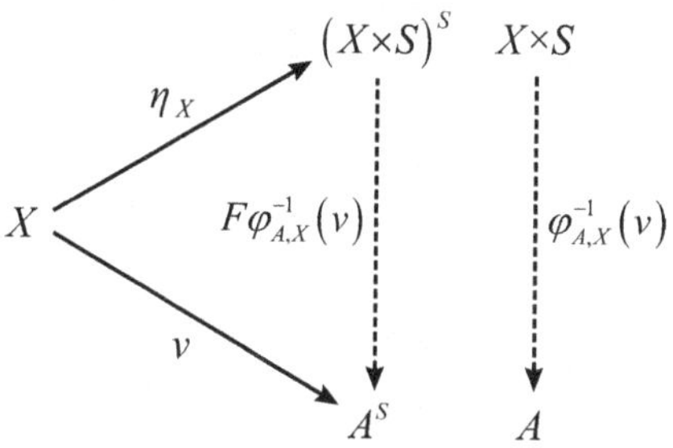

This, of course, is a statement of the universal property that for each set X, there exists an initial morphism from X to F (cf. Definition 10.13). The element chase version of the commutative diagram (17) is

(18)

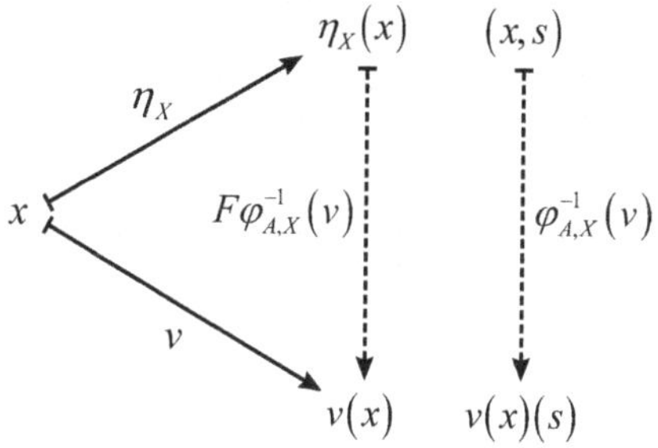

By definition of the hom-functor F,

$$\left(F\varphi_{A,X}^{-1}(v)\right)(\eta_X(x)) = \varphi_{A,X}^{-1}(v) \circ \eta_X(x); \tag{19}$$

when evaluated at $s \in S$,

$$\begin{aligned} \left\{\varphi_{A,X}^{-1}(v) \circ \eta_X(x)\right\}(s) &= \left(\varphi_{A,X}^{-1}(v)\right)(\eta_X(x)(s)) \\ &= \left(\varphi_{A,X}^{-1}(v)\right)(x, s) \\ &= (v(x))(s). \end{aligned} \tag{20}$$

This reconciles the two components of diagram (18).

11.5 Counit and Terminal Morphism The component of the counit of this adjunction at set A is

(21) $$\bar{\eta}_A = \varphi^{-1}_{A,A^S}(1_{A^S}) \in \mathbf{Set}(A^S \times S, A),$$

which is the *evaluation mapping* $e : A^S \times S \to A$ (cf. Section 0.24), defined, for $k : S \to A$ and $s \in S$, by

(22) $$e(k,s) = \bar{\eta}_A(k,s) = \left(\varphi^{-1}_{A,A^S}(1_{A^S})\right)(k,s) = k(s).$$

When expressed involving the evaluation mapping, the sequential evaluation (8) is, for $x \in X$ and $s \in S$,

(23) $$\begin{aligned} e\big((\varphi_{A,X}(u))(x),s\big) &= \bar{\eta}_A\big((\varphi_{A,X}(u))(x),s\big) \\ &= \{(\varphi_{A,X}(u))(x)\}(s) = u(x,s) \end{aligned}.$$

The unfolding of the material entailment (7) into the hierarchical composition (9), the essence of the adjunction isomorphism φ (6), may then be represented in the commutative diagram

(24)

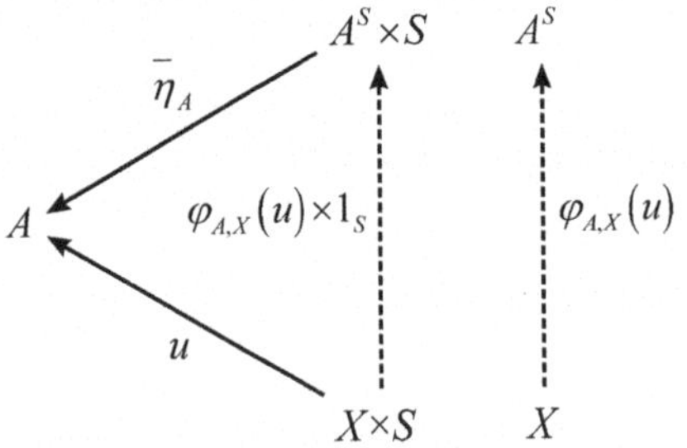

This diagram illustrates that the hom-set $FA = A^S = \mathbf{Set}(S,A)$ and the counit component mapping $\bar{\eta}_A \in \mathbf{Set}(GF\,A, A) = \mathbf{Set}(A^S \times S, A)$ are such that, for any mapping $u : X \times S \to A$, there is a unique mapping $\varphi_{A,X}(u) : X \to A^S$ satisfying $\bar{\eta}_A \circ G\varphi_{A,X}(u) = \bar{\eta}_A \circ (\varphi_{A,X}(u) \times 1_S) = u$. This, dually, is a statement of the universal property that for each set A, there exists a terminal morphism from G to A (cf. Definition 10.13). The element chase version of the commutative diagram (24) is

(25)

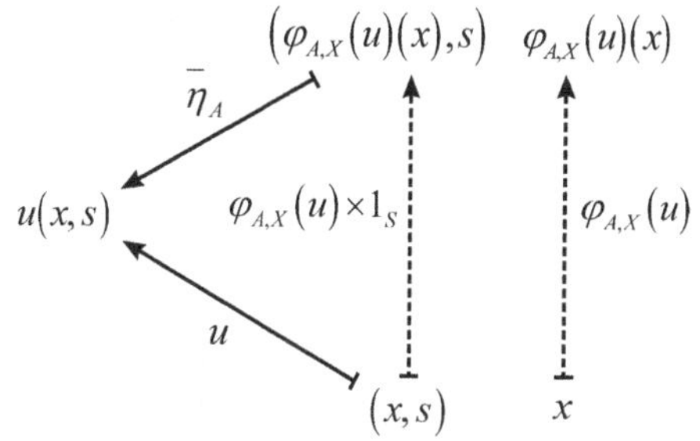

11.6 Counit–Unit Equations The first counit–unit equation $F\bar{\eta} \circ \eta_F = 1_F$ (cf. Section 10.5) for the present example is an equality of natural transformations as morphisms in the functor category (Definition 0.19) $\mathbf{Set}^{\mathbf{Set}}$. In component form (cf. Section 10.4), for each set A, since $FA = A^S$, the equation is

(26) $$F\bar{\eta}_A \circ \eta_{A^S} = 1_{A^S} \in \mathbf{Set}(A^S, A^S).$$

The left-hand side is

(27) $$\eta_{A^S} : A^S \to (A^S \times S)^S$$

sequentially composed with

(28) $$F\bar{\eta}_A : (A^S \times S)^S \to A^S.$$

For $h \in A^S$, i.e., mapping $h : S \to A$, and $s \in S$, the relay is

(29) $$\eta_{A^S} : h \mapsto [s \mapsto (h, s)];$$

followed by

(30) $$F\bar{\eta}_A : [s \mapsto (h, s)] \mapsto [s \mapsto h(s)] = h.$$

This verifies $F\bar{\eta}_A \circ \eta_{A^S}(h) = h = 1_{A^S}(h)$, hence equality (26).

The second counit–unit equation $\bar{\eta}_G \circ G\eta = 1_G$ (in $\mathbf{Set}^{\mathbf{Set}}$) in component form is, for each set X, since $GX = X \times S$,

(31) $$\bar{\eta}_{X\times S} \circ G\eta_X = 1_{X\times S} \in \mathbf{Set}(X \times S, X \times S).$$

The left-hand side is

(32) $$G\eta_X : X \times S \to (X \times S)^S \times S$$

sequentially composed with

(33) $$\bar{\eta}_{X\times S} : (X \times S)^S \times S \to X \times S.$$

For $(x, s) \in X \times S$, the relay is

(34) $$G\eta_X : (x, s) \mapsto ([\bullet \mapsto (x, \bullet)], s);$$

followed by

(35) $$\bar{\eta}_{X\times S} : ([\bullet \mapsto (x, \bullet)], s) \mapsto (x, s).$$

This verifies $\bar{\eta}_{X\times S} \circ G\eta_X(x, s) = (x, s) = 1_{X\times S}(x, s)$, hence equality (31).

11.7 Cartesian Closed Category The **Set**-isomorphisms (6), or more generally

(36) $$\varphi : \mathbf{C}(X \times S, A) \cong \mathbf{C}(X, A^S),$$

are involved in the general definitions of exponentials, Cartesian closed category, hierarchical composition, and functional entailment (*ML*: A.52, A.53; *RL*: 6.12–6.14). This adjunction

(37) $$\bullet \times S \dashv (\bullet)^S$$

has myriad implications, particularly in the ontogenesis of (M, R)-systems (recall that replication depends on the existence of an inverse evaluation map, Section 8.5; see also *ML*: 11.15–11.17 and 13.23–13.27) and in therapeutics (*RL*: 14.9–14.12) .

Galois Connection

> Sauter à pieds joints sur ces calculs, grouper les opérations, les classer suivant leurs difficultés et non suivant leurs formes; telle est, suivant moi, la mission des géomètres futurs; telle est la voie où je suis entré dans cet ouvrage.
>
> [Go to the roots of these calculations! Group the operations. Classify them according to their complexities rather than their appearances! This, I believe, is the mission of future mathematicians. This is the road on which I am embarking in this work.]
>
> — Évariste Galois (1831)
> Préface pour «*Deux Mémoires d'Analyse pur*»
> December 1831

11.8 Thin Categories and Isotone Functors In Sections 4.19 and 4.20, I explicated the categorical connections of preordered [or partial order] sets, that a proset [or poset] $\langle X, \leq \rangle$ is a thin category $\mathbf{C}$, that $\mathfrak{O}\mathbf{C} = X$ and $\mathfrak{A}\mathbf{C} = \leq$, and that an isotone mapping (Definition 4.13), a preorder–preserving [or partial-order–preserving] morphism, is a functor between thin categories.

For example, given a mapping $f : Y \to Z$, the inverse power set mapping $\overline{\mathrm{P}}f : \langle \mathrm{P}Z, \subset \rangle \to \langle \mathrm{P}Y, \subset \rangle$ is an isotone mapping from the poset $\langle \mathrm{P}Z, \subset \rangle$ to the poset $\langle \mathrm{P}Y, \subset \rangle$. It is therefore the functor

$$\overline{\mathrm{P}}f : \begin{cases} A \mapsto f(A) & (A \in \mathrm{P}Z) \\ [A \subset B] \mapsto [f^{-1}(A) \subset f^{-1}(B)] & (A, B \in \mathrm{P}Z) \end{cases}. \tag{38}$$

Note that (38) shows the object and arrow actions of the isotone mapping $\overline{\mathrm{P}}f$ as a *covariant* functor from the category $\langle \mathrm{P}Z, \subset \rangle$ to the category $\langle \mathrm{P}Y, \subset \rangle$. These are distinct from the actions of $\overline{\mathrm{P}} : \mathbf{Set} \to \mathbf{Set}$ as a *contravariant* functor (cf. Definition 3.22).

When specialized to thin categories, adjunction becomes a particular correspondence between prosets [/posets] called Galois connection, which appears in many branches of mathematics.

11.9 Isotone Galois connection Let $\langle A, \leq_A \rangle$ and $\langle X, \leq_X \rangle$ be two p[r]osets. An *isotone Galois connection* from $\langle X, \leq_X \rangle$ to $\langle A, \leq_A \rangle$ consists of two isotone mappings $G : X \to A$ and $F : A \to X$, such that for all $a \in A$ and $x \in X$,

(39) $$G(x) \leq_A a \quad \text{if and only if} \quad x \leq_X F(a).$$

The isotone mapping G is called a *left adjoint* of F, F is called a *right adjoint* of G, and their connection is denoted by

(40) $$G \dashv F.$$

This definition is, of course, simply a special case of the general category-theoretic hom-set Definition 10.2, when the p[r]osets and isotone mappings involved are considered as their corresponding categories and functors:

(41) $$\langle A, \leq_A \rangle \underset{F}{\overset{G}{\leftrightarrows}} \langle X, \leq_X \rangle.$$

In a Galois connection, due to the positions of the mappings relative to the orders $\leq$ in (39), the left adjoint G is also called the 'lower adjoint', and the right adjoint F the 'upper adjoint'.

11.10 Closure and Interior Recall (Definition 4.21) that for two mappings $f, g : X \to Z$ with the same proset $\langle Z, \leq \rangle$ as codomain is imputed with the preorder $f \preceq g$ when for all $x \in X\, f(x) \leq g(x)$.

A mapping $P : Z \to Z$ from a p[r]oset $\langle Z, \leq \rangle$ to itself is

(i) *idempotent* if $P^2 = P$, i.e., for all $x \in X\; P(P(x)) = P(x)$, which means that repeated application does not alter the output of the initial application;

(e) *extensive* if $1_X \preceq P$, i.e., for all $x \in X\; x \leq P(x)$;

(c) *constrictive* if $P \preceq 1_X$, i.e., for all $x \in X\; P(x) \leq x$;

($^-$) a *closure operator* if it is extensive, idempotent, and isotone (cf. *RL*: 3.24);

($^\circ$) an *interior operator* if it is constrictive, idempotent, and isotone.

The usage of 'closure' and 'interior' originated in topology, in which the closure $\overline{A}$ of a set A in a topological space is the union of A and the set of its accumulation points, and the interior A° of A is the complement of the closure of the complement of A, $A^\circ = \left(\overline{A^c}\right)^c$, whence closure and interior are dual notions.

Given a Galois connection with left adjoint G and right adjoint F, the composition $F \circ G : X \to X$ is called the *associated closure operator*, and $G \circ F$:

$A \to A$ is called the *associated interior operator*. Both mappings are isotone and idempotent, and

$$(42) \qquad \begin{array}{llr} (e) & \text{for all } x \in X & x \leq F \circ G(x) \\ (c) & \text{for all } a \in A & G \circ F(a) \leq a \end{array}.$$

11.11 Counit–Unit Definition Note the form of inequalities (42); they say

$$(43) \qquad \begin{array}{llll} (e) & 1_X & \preceq F \circ G & (\text{in } \mathbf{Pro}\ (X,X)) \\ (c) & G \circ F & \preceq 1_A & (\text{in } \mathbf{Pro}\ (A,A)) \end{array},$$

which precisely correspond to the counit–unit definition of adjunction (Definition 10.4). Recall (Section 4.22) that $\mathbf{Pro}(X,Z)$ may be considered as a functor category, whence for $f, g \in \mathbf{Pro}(X,Z) = Z^X$ there is a natural transformation from the functor f to the functor g if and only if $f \preceq g$. Thus the preorder relations in (43) precisely define the requisite natural transformations that are the unit and the counit.

11.12 Uniqueness and Invertibility Let the isotone mappings $G_1, G_2 : \langle X, \leq_X \rangle \to \langle A, \leq_A \rangle$ and $F_1, F_2 : \langle A, \leq_A \rangle \to \langle X, \leq_X \rangle$ between prosets form adjunctions $G_1 \dashv F_1$ and $G_2 \dashv F_2$. Then $F_1 \preceq F_2$ if and only if $G_1 \preceq G_2$, whence $F_1 \cong F_2$ if and only if $G_1 \cong G_2$. Thus the adjoint of an isotone mapping is unique up to **Pro**-isomorphism if it exists.

Since a partial order is a preorder for which the relation $\cong$ coincides with = (Section 4.12), if $\langle A, \leq_A \rangle$ and $\langle X, \leq_X \rangle$ are posets, then the adjoints are unique if they exist. In a Galois connection of posets, therefore, a left/right adjoint uniquely determines the other:

$$(44) \qquad \begin{array}{l} G(x) = \inf\{a \in \langle A, \leq_A \rangle : x \leq_X F(a)\} \\ F(a) = \sup\{x \in \langle X, \leq_X \rangle : G(x) \leq_A a\} \end{array}.$$

The extremalities imply that if G or F is invertible, then the pair are mutual inverses, i.e., $F = G^{-1}$ and $G = F^{-1}$.

11.13 Antitone Galois connection Let $\langle A, \leq_A \rangle$ and $\langle X, \leq_X \rangle$ be two p[r]osets. An *antitone Galois connection* between these p[r]osets consists of two antitone mappings (Definition 4.16) $G : X \to A$ and $F : A \to X$, such that for all $a \in A$ and $x \in X$,

(45) $$a \leq_A G(x) \text{ if and only if } x \leq_X F(a).$$

In an antitone Galois connection, the roles of F and G, and that of the p[r]osets $\langle A, \leq_A \rangle$ and $\langle X, \leq_X \rangle$, are symmetric. The loss of chirality means the 'adjoints' are ambidextrous, and the two mappings are then called *polarities*. Also, an antitone Galois connection is simply *between* two p[r]osets, rather than *from* one *to* the other.

As is the case for isotone Galois connection, when $\langle A, \leq_A \rangle$ and $\langle X, \leq_X \rangle$ are posets, each polarity still uniquely determines the other, with

(46) $$\begin{aligned} G(x) &= \sup\{a \in \langle A, \leq_A \rangle : x \leq_X F(a)\} \\ F(a) &= \sup\{x \in \langle X, \leq_X \rangle : a \leq_A G(x)\} \end{aligned}.$$

The composites $F \circ G : X \to X$ and $G \circ F : A \to A$ are *both* associated *closure* operators. They are isotone, (*i*) idempotent, and

(47) $$\begin{aligned} &(e) \quad \text{for all } x \in X \quad x \leq F \circ G(x) \\ &(e) \quad \text{for all a} \in A \quad a \leq G \circ F(a) \end{aligned}.$$

The implications of the two definitions of Galois connections are very similar (indeed, *dual* in a precise sense), since an antitone Galois connection between $\langle A, \leq_A \rangle$ and $\langle X, \leq_X \rangle$ is equivalently an isotone Galois connection from $\langle A, \leq_A \rangle$ to the dual p[r]oset $\widetilde{X} = \langle X, \geq_X \rangle$ of $\langle X, \leq_X \rangle$ (cf. Definition 4.12). Thus all the statements and examples below in one of the two Galois connections have corresponding duals in the other. Antitone being 'isotone to the dual proset' is the specialization to prosets of a general situation, that contravariance is 'covariance to the opposite category'. Contravariant functors (Definition 0.11) often get mentioned in passing, but do not see much actual action in the trenches of category theory. It is mainly because a contravariant functor $F : \mathbf{C} \to \mathbf{D}$ is a covariant functor $F : \mathbf{C} \to \mathbf{D}^{op}$ (*ML*: A.10), where $\mathbf{D}^{op}$ is the opposite category of $\mathbf{D}$ (*ML*: A.8), formed by 'reversing all the arrows'. Explicitly, $\mathfrak{O}\mathbf{D}^{op} = \mathfrak{O}\mathbf{D}$, and to each $f \in \mathbf{D}(X, Y) \subset \mathfrak{A}\mathbf{D}$ there corresponds an $f^{op} \in \mathbf{D}^{op}(Y, X) \subset \mathfrak{A}\mathbf{D}^{op}$.

The general correspondence that is Galois connection was introduced by Øystein Ore in the now-classic paper [Ore 1944], and the name was taken from the original Galois theory of equations, where the correspondence between subgroups and subfields represented an antitone Galois connection.

11.14 Galois theory Galois theory deals with field extensions and automorphism groups. Ore's motivation was from a theorem therein called the Fundamental Galois Pairing. I will only give its statement here without dwelling into all the associated concepts. The interested reader is referred to the very readable, introductory treatment Hadlock [1978].

Let E be a normal extension of the field D, and let X be the poset of all subfields of E that contain D, ordered by set inclusion. For $K \in X$, the group of field automorphisms of E that hold K fixed is called the Galois group of E over K, denoted $\mathrm{Gal}(E/K)$. Let A be the poset of all subgroups of $\mathrm{Gal}(E/D)$, ordered by set inclusion. For $H \in A$, let $\mathrm{Fix}(H)$ be its fixed field, the field of elements of E held fixed by the group members of H. Then the maps $G : X \to A$ and $F : A \to X$, defined by

(48)
$$\begin{aligned} G &: K \mapsto \mathrm{Gal}(E/K) \\ F &: H \mapsto \mathrm{Fix}(H) \end{aligned}$$

form an antitone Galois connection. The correspondence is order-reversing: $K_1 \subset K_2$ iff $\mathrm{Gal}(E/K_1) \supset \mathrm{Gal}(E/K_2)$.

In relational biology, a Galois theory of equivalence relation lattices and bijection groups on a set has also been developed. The enthused reader may consult Section IV.F.20 in Louie [1985].

11.15 Power Set Algebra Consider the power set algebra $\mathrm{P}Z$ and fix a subset $Y \subset Z$. Then the mappings $G, F : \langle \mathrm{P}Z, \subset \rangle \to \langle \mathrm{P}Z, \subset \rangle$, defined by

(49)
$$\begin{aligned} G(X) &= Y \cap X \quad (X \subset Z) \\ F(A) &= A \cup Y^c \quad (A \subset Z) \end{aligned},$$

form an isotone Galois connection, with the defining biconditional being

(50)
$$Y \cap X \subset A \quad \text{if and only if} \quad X \subset A \cup Y^c.$$

A similar Galois connection can be found in any Boolean algebra (*ML*: 3.9–3.20). Explicitly, the two isotone mappings are

(51)
$$\begin{aligned} G(x) &= y \wedge x \\ F(a) &= a \vee \neg y = [y \to a] \end{aligned}.$$

The defining biconditional of their isotone Galois connection is

$$(52) \qquad y \wedge x \rightarrow a \quad \text{if and only if} \quad x \rightarrow [a \vee \neg y],$$

both statements reducing to $a \vee \neg y \vee \neg x$. In logical terms, 'implication from y' is the right adjoint of 'conjunction with y'.

11.16 Left and Right *R*-Relatives Let $R \in \mathbf{Rel}(Y, Z)$. For a subset $X \in \mathrm{P}Y$, define $G(X)$ to be the range of the restriction $R|_X$ of R to X (cf. Definitions 1.18 and 1.26); it is the set of all right R-relatives of elements of X:

$$(53) \qquad \begin{aligned} G(X) &= \operatorname{ran}\left(R|_Z\right) \\ &= \{z \in Z : \exists\, x \in X (x, z) \in R\} \subset \operatorname{ran}(R) \subset Z \end{aligned}.$$

Correspondingly, for a subset $A \in \mathrm{P}Z$, define $F(A)$ to be the corange of the restriction $R|^A$ of R to A; it is the set of all left R-relatives of elements of A:

$$(54) \qquad \begin{aligned} F(A) &= \operatorname{cor}\left(R|^A\right) \\ &= \{y \in Y : \exists\, a \in A (a, y) \in R\} \subset \operatorname{cor}(R) \subset Y \end{aligned}.$$

Then $G : \langle \mathrm{P}Y, \subset \rangle \rightarrow \langle \mathrm{P}Z, \subset \rangle$ and $F : \langle \mathrm{P}Z, \subset \rangle \rightarrow \langle \mathrm{P}Y, \subset \rangle$ form an antitone Galois connection.

11.17 Image and Inverse Image Let $f \in \mathbf{Set}(Y, Z)$. For a subset $X \in \mathrm{P}Y$, define $G(X)$ to be the image

$$(55) \qquad G(X) = \mathrm{P}f(X) = f(X) = \{f(x) \in Z : x \in X\} \subset Z.$$

For a subset $A \in \mathrm{P}Z$, define $F(A)$ to be the inverse image

$$(56) \qquad F(A) = \overline{\mathrm{P}}f(A) = f^{-1}(A) = \{x \in Y : f(x) \in A\} \subset Y.$$

Then $G: \langle \mathrm{P}Y, \subset \rangle \to \langle \mathrm{P}Z, \subset \rangle$ is the left adjoint and $F: \langle \mathrm{P}Z, \subset \rangle \to \langle \mathrm{P}Y, \subset \rangle$ is the right adjoint of an isotone Galois connection. The defining biconditional is

$$(57) \qquad f(X) \subset A \quad \text{if and only if} \quad X \subset f^{-1}(A).$$

11.18 Inverse Image and Dual Image Further, for a subset $X \in \mathrm{P}Y$, one may define a 'dual image'

$$(58) \qquad H(X) = \left\{ z \in Z : f^{-1}(z) \subset X \right\} \subset Z.$$

Note that if $z \notin \mathrm{ran}(f)$, i.e., $z \in Z \sim \mathrm{ran}(f)$, then $f^{-1}(z) = \varnothing \subset X$; so for all $X \in \mathrm{P}Y$, $Z \sim \mathrm{ran}(f) \subset H(X)$. The image (55) can also be represented as

$$(59) \qquad \begin{aligned} G(X) &= \mathrm{P}f(X) = f(X) \\ &= \{z \in \mathrm{ran}(f) : \exists\, x \in X\ [f(x) = z]\} \\ &= \left\{ z \in \mathrm{ran}(f) : f^{-1}(z) \subset X \right\} \subset Z, \end{aligned}$$

whence $G(X) = \mathrm{ran}(f) \cap H(X)$. Thus

$$(60) \qquad \begin{aligned} H(X) = Z \cap H(X) &= \{\mathrm{ran}(f) \cup [Z \sim \mathrm{ran}(f)]\} \cap H(X) \\ &= \{\mathrm{ran}(f) \cap H(X)\} \cup \{[Z \sim \mathrm{ran}(f)] \cap H(X)\} \\ &= G(X) \cup [Z \sim \mathrm{ran}(f)]. \end{aligned}$$

Consider the inclusion $f^{-1}(A) \subset X$ for $A \in \mathrm{P}Z$ and $X \in \mathrm{P}Y$. One may likewise partition A into the disjoint union $A = \{\mathrm{ran}\,(f) \cap A\} \cup \{[Z \sim \mathrm{ran}\,(f)] \cap A\}$. One has $f^{-1}([Z \sim \mathrm{ran}(f)] \cap A) = \varnothing$, and one may readily verify that therefore

$$(61) \qquad \begin{aligned} f^{-1}(A) \subset X \quad &\Leftrightarrow \quad f^{-1}(\mathrm{ran}(f) \cap A) \subset X \\ &\Leftrightarrow \quad \mathrm{ran}(f) \cap A \subset f(X) \end{aligned}.$$

The biconditional

$$(62) \qquad F(A) \subset X \quad \text{if and only if} \quad A \subset H(X)$$

that defines an isotone Galois connection with left adjoint $F : \langle \wp Z, \subset \rangle \to \langle \wp Y, \subset \rangle$ and right adjoint $H : \langle \wp Y, \subset \rangle \to \langle \wp Z, \subset \rangle$, in view of (60) and (61), then becomes

$$\text{(63)} \qquad \operatorname{ran}(f) \cap A \subset f(X) \quad \text{if and only if} \quad A \subset f(X) \cup [Z \sim \operatorname{ran}(f)],$$

which is now isomorphic to the logical statement (52).

Note that in the first Galois connection in Section 11.17, F is the right adjoint, while in the second Galois connection in Section 11.18, F is the left adjoint. Stated otherwise, the isotone inverse-image mapping $F = \overline{\wp} f = f^{-1}$ between power sets possesses both left and right adjoints, and the two adjoints are different.

11.19 Quantifiers as Adjoints The image (55) may alternatively be denoted

$$\text{(64)} \qquad \begin{aligned} G(X) &= \wp f(X) = f(X) \\ &= \{z \in Z : \exists x \in Y\, [f(x) = z \wedge x \in X]\} \subset Z \end{aligned}.$$

The dual image (58) may alternatively be denoted

$$\text{(65)} \qquad \begin{aligned} H(X) &= \{z \in Z : f^{-1}(z) \subset X\} \\ &= \{z \in Z : \forall x \in Y\, [f(x) = z \to x \in X]\} \subset Z \end{aligned}.$$

Thus, given an $f \in \mathbf{Set}(Y, Z)$, the left adjoint of the isotone inverse-image mapping $F = \overline{\wp} f = f^{-1}$ is the *existential quantifier along f*, and may be denoted $G = \exists_f$:

$$\text{(66)} \qquad \wp Z \underset{f^{-1}}{\overset{\exists_f}{\leftrightarrows}} \wp Y.$$

Dually, the right adjoint of the isotone inverse-image mapping $F = \overline{\wp} f = f^{-1}$ is the *universal quantifier along f*, and may be denoted $H = \forall_f$:

$$\text{(67)} \qquad \wp Y \underset{\forall_f}{\overset{f^{-1}}{\leftrightarrows}} \wp Z.$$

The connections to quantifiers is more formally evident when one lets the mapping f be the canonical projection $\pi_1 : Y \times Z \to Y$ (Definition 1.16), where $\pi_1(y,z) = y$. Then, for $A \in \mathrm{P}Y$ and $X \in \mathrm{P}(Y \times Z)$,

$$(68)\qquad \begin{aligned} F(A) &= \pi_1^{-1}(A) = A \times Z \in \mathrm{P}(Y \times Z) \\ G(X) &= \exists_{\pi_1}(X) = \{y \in Y : \exists z \in Z\ [(y,z) \in X]\} \in \mathrm{P}Y \\ H(X) &= \forall_{\pi_1}(X) = \{y \in Y : \forall z \in Z\ [(y,z) \in X]\} \in \mathrm{P}Y. \end{aligned}$$

11.20 Semantics–Syntax Adjunction I mention *en passant* Lawvere's final observation [Lawvere 1969] that in the foundations of mathematical logic there is an antitone Galois connection between sets of axioms and classes of models. Let X be the collection of all logical theories (axiomatizations), and A be the power set of the collection of all mathematical structures. For a theory T in X, let $G(T)$ be the set of all structures that satisfy the axioms (whence are models of) T. For a set of mathematical structures $S \subset A$, let $F(S)$ be the minimal axiomatization of S. Then

$$(69)\qquad G(T) \subset S \quad \text{if and only if} \quad T \to F(S).$$

This says the 'semantics functor' G is left adjoint to the 'syntax functor' F in the isotone Galois connection

$$(70)\qquad \widetilde{X} = \text{theories}^{op} \underset{F=\text{syntax}}{\overset{G=\text{semantics}}{\rightleftarrows}} A = \text{structures}.$$

Or, equivalently, that syntax and semantics are adjoints (polarities) in an antitone Galois connection.

12
Free and Forgetful

النسيان شكل من أشكال الحرية

[Forgetfulness is a form of freedom.]

— Kahlil Gibran (1926)
Sand and Foam

Algebraic Structures

12.1 *n*-ary Operation An *n-ary operation* λ *on a* (nonempty) *set* S is a (single-valued) mapping $\lambda : S^n \to S$ (whence $\lambda \subset S^n \times S = S^{n+1}$). The number n, called the *arity* of the operation, is usually a (finite) nonnegative integer, but may be naturally extended to be infinite, countable or uncountable.

When $n = 0$, $S^0 = \{0\} = 1$ (cf. Section 2.4); a *nullary operation*, $\lambda : 1 \to S$, is the postulate of the existence of a special element $\lambda(0) \in S$. A *unary operation* (when $n = 1$, also called an *operator*), $\lambda : S \to S$, is a standard mapping that is a single-valued relation on S. One often encounters *binary operations* $\lambda : S \times S \to S$, when $n = 2$ (cf. Section 0.22).

The general notion of an *algebraic structure* is set S with nullary, unary, binary, ternary, … operations satisfying as axioms a variety of identities between composite operations. The subject of *universal algebra* is concerned with the general properties of such a structure.

12.2 Group A *group* $\left\langle S, \cdot, e, (\bullet)^{-1} \right\rangle$ is an algebraic structure consisting of a set S together with

i. a binary operation $\cdot : S \times S \to S$, often called the *product*, and denoted multiplicatively: for $x, y \in S$, one writes

A. H. Louie, *Intangible Life*, Anticipation Science 2,
DOI 10.1007/978-3-319-65409-6_13

(1) $$\cdot\,(x,y) = x \cdot y,$$

or simply by juxtaposition

(2) $$\cdot\,(x,y) = xy;$$

ii. a nullary operation $e : 1 \to S$, for which the special element $e = e(1) \in S$ is called the *identity*, and
iii. a unary operation $(\bullet)^{-1} : S \to S$, such that for $x \in S$ the element $x^{-1} = (\bullet)^{-1}(x) \in S$ is called the *inverse* of x.

These operations on S satisfy the following properties:

(g1) the binary operation $\cdot$ is *associative*; i.e., for all $x, y, z \in S$,

(3) $$(x \cdot y) \cdot z = x \cdot (y \cdot z);$$

(g2) the identity element e is such that, for every element $x \in S$,

(4) $$x \cdot e = e \cdot x = x;$$

(g3) for every element $x \in S$, there exists an inverse element $x^{-1} \in S$ such that

(5) $$x \cdot x^{-1} = x^{-1} \cdot x = e.$$

The group concept may be generalized by stepwise removal of axioms. Deleting the unary inverse operation iii [whence also the property (g3)] and one obtains the algebraic structure of monoid. Explicitly,

12.3 Monoid A *monoid* is an algebraic structure consisting of a set together with

i. an associative binary operation, and
ii. an identity element.

Dropping also the nullary operation ii that specifies the identity [whence also the property (g2)] gives a semigroup.

12.4 Semigroup A *semigroup* is an algebraic structure consisting of a set together with

i. an associative binary operation.

In anabolic synthesis instead of catabolic analysis, a monoid is a semigroup with an identity element, and a group is a monoid further equipped with inverses.

The set $\mathbb{N} = \{1, 2, 3, 4, \ldots\}$ of natural numbers (positive integers) with the addition operation is a semigroup but not a monoid. $\mathbb{N}$ with the multiplication operation is a monoid but not a group (the element 1 being the identity). The set $\mathbb{N}_0 = \{0\} \cup \mathbb{N} = \{0, 1, 2, 3, 4, \ldots\}$ of whole numbers (nonnegative integers) with the addition operation is also a monoid but not a group (the element 0 being the identity). The set $\mathbb{Z} = \{\ldots, -3, -2, -1, 0, 1, 2, 3, \ldots\}$ of integers with the addition operation is a group (the inverse of $n \in \mathbb{Z}$ being $-n \in \mathbb{Z}$).

12.5 Categorical Connections A monoid is a category with just one single object. The arrows of the category are the elements of the monoid, and the identity arrow is the identity element. Composition of arrows is the binary operation of the monoid.

More generally, for any object A in any category $\mathbf{C}$, the set of arrows from A to A, i.e., the hom-set $\mathbf{C}(A, A)$, is a monoid under the composition operation of $\mathbf{C}$. (See *RL*: 3.20–3.22 for an exposition of the monoid $\mathbf{Rel}(X, X)$.)

On account of the existence of inverses, a group is a single-object category in which every arrow is an isomorphism. (Compare property (g3) with Definition 0.5.)

Monoids, just like other algebraic structures, form their own category. **Mon** is the category in which objects are monoids and morphisms are monoid homomorphisms, which are mappings that preserve the monoid structure: $f \in \mathbf{Mon}(A, B)$ must satisfy the properties $f(a \cdot_A b) = f(a) \cdot_B f(b)$ and $f(e_A) = e_B$. Monoid homomorphisms are, indeed, functors between single-object categories.

Free Objects

12.6 Word Given a set X, a *word* over X is a finite sequence (or string) of elements of X. The number of elements in the finite sequence is the *length* of the word. The *empty word* e is the word with no elements at all (i.e., a string of length zero).

12.7 Free Monoid The *free monoid* over the set X is defined to be the set

(6) $$X^* = \{\text{words over} X\}.$$

The product of two words is their concatenation. Members of X are called *generators*, and if X is a finite set, X^* is called *finitely generated*. Since a string of generators is called a word, the set X has also acquired the name of 'alphabet', and its members 'letters'. Evidently, each $x \in X$ is a word (of length 1) over X, whence $X \subset X^*$.

A free monoid is 'free' in the sense that it is subject to no *nontrivial* relations; that is, there are no binding relations among the words other than those required by the monoid axioms ($g1$) associativity and ($g2$) identity, i.e., equations (3) and (4).

More formally, the free monoid over the set X is constructed thus: given a set X, define $X_0 = \{e\}$ (the singleton set consisting solely of the empty word), $X_1 = X$, and then recursively, for each $i > 0$,

(7) $$X_{i+1} = \{yx : y \in X_i, x \in X\}.$$

X_i is therefore the set of all words of length i, i.e., the concatenation of i letters in X, an 'i-letter word'. Then define the free monoid over X as

(8) $$X^* = \bigcup_{i \in \mathbb{N}_0} X_i = \{e\} \cup X \cup X_2 \cup X_3 \cup X_4 \cup \cdots.$$

(A related structure is the *free semigroup*, which is all *nonempty* words over X, i.e., the set $X^+ = \bigcup_{i \in \mathbb{N}} X_i$ with concatenation.)

12.8 Infinitude If X is the empty set $\varnothing$, then $X^* = \{e\}$. If X is a countable set, finite or infinite, the free monoid X^* is a countably infinite set. The inclusion $X \subset X^*$ implies that if X is uncountably infinite, then so is X^*.

One may note the degenerate case of a singleton set X in which the sole element is the empty word e; then one would have $X^* = \{e\}$. But this X violates the 'no nontrivial relations' requirement of a free monoid, since each $x \in X$ must satisfy $x = e$ (a nontrivial equation which is not a requisite monoid axiom). So $X = \{e\}$ cannot be a generator set.

The countably infinite set $\mathbb{N}_0 = \{0, 1, 2, 3, 4, \ldots\}$ of whole numbers with the addition operation is, for example, a free monoid with the singleton generator set $X = \{1\}$. According to the formal definition, $X_0 = \{e\} = \{0\}$, $X_1 = X = \{1\}$,

$X_2 = \{1+1\} = \{2\}, X_3 = \{2+1\} = \{3\}$, etc., whence $X^* = \mathbb{N}_0 = \{0, 1, 2, 3, 4, \ldots\}$. The *Fundamental Theorem of Arithmetic* states that the monoid $\mathbb{N}$ of natural numbers (positive integers) under multiplication is a free monoid over the countably infinite set of generators, the prime numbers.

Two free monoids X^* and Y^* are isomorphic if and only if X and Y have the same cardinality. This cardinality of the set of generators is called the *rank* of the free monoid. A free monoid is finitely generated if and only if it has finite rank. For every cardinal number n, there is, up to isomorphism, exactly one free monoid of rank n, which may be represented canonically as n^*. The examples in the previous paragraph show $\langle \mathbb{N}_0, + \rangle \cong 1^*$ and $\langle \mathbb{N}, \times \rangle \cong \aleph_0^*$. The free monoid $2^* = \{0, 1\}^*$ (with concatenation) contains all possible finite strings of 0 and 1s, and is, therefore, metaphorically as well as literally the collection of all possible algorithms.

12.9 Verbum Caro Factum Est The set of all nucleotide sequences is a free monoid over the mRNA 4-letter alphabet $R = \{\text{U, C, A, G}\}$ (or, equivalently, the DNA alphabet $\{\text{T, C, A, G}\}$). Or better, in terms of the nucleotide triplets that are codons, the generator set is the alphabet of $4^3 = 64$ letters,

$$(9) \qquad X = R^3\{\text{UUU, UUC, UUA, UUG, AUU}, \ldots, \text{GGA, GGG}\},$$

and the countably infinite set of all possible nucleotide sequences is the finitely generated free monoid X^*.

A protein molecule is a word over the 20-letter alphabet of amino acids. The countably infinite set of all polypeptide sequences (presumably only a finite subset of which constitutes viable proteins) is then the free monoid M^* over the 20-letter alphabet of $M = \{20 \text{ amino acids}\}$.

The genetic code is, therefore, a monoid homomorphism γ from $X^* = 64^*$ to $M^* = 20^*$. (The actual situation is slightly more complicated, because of the 'Start' and 'Stop' codes in translation. The most common start codon is AUG, which is read as the amino acid methionine; but alternative start codons, depending on the organism, include GUG or UUG. There are three stop codons, UAG, UGA, and UAA, which do not code for any amino acids, and, instead, signal the release of the nascent polypeptide from the ribosome. One may consider the domain of $\gamma : X^* \to M^*$ to be $X^* = 61^*$, and a posteriori 'bracket' a polypeptide word $\in M^*$ by appending 'Start' and 'Stop' to its initial and terminal ends.)

12.10 Universal Monoid Given any mapping g from a set X to (the underlying set of) a monoid $\langle M, \cdot \rangle$, there exists a unique monoid homomorphism $f : X^* \to M$ that makes the following diagram commute:

(10) 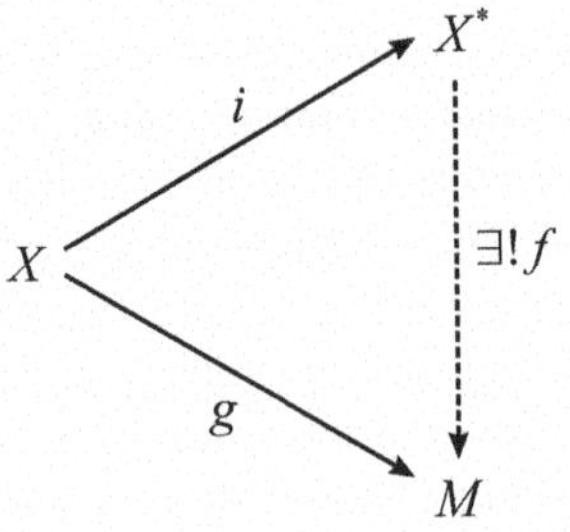

where $i : X \to X^*$ is the inclusion map from X into X^* (cf. Definition 1.2).

One may construct f explicitly. First, f must send the empty word $e \in X^*$ to the identity element of M. The inclusion map $i : X \to X^*$ sends each symbol x in X to the one-letter word $i(x) = x$ in X^* consisting of that symbol. f has to agree with g on the elements of X, so one must define $f(i(x)) = g(x)$. For the remaining words in X^* (consisting of strings of more than one symbol), f may be uniquely extended since it is a monoid homomorphism; i.e., if $w = x_1 x_2 \cdots x_k \in X^*$, then $f(w) = g(x_1) \cdot g(x_2) \cdot \cdots \cdot g(x_k)$. One concludes that the action of a monoid homomorpism with a free monoid X^* as domain is uniquely determined by its values on the generator set X.

12.11 Free Group Given a set X, first define formally for each $x \in X$ an *inverse* x^{-1}. The collection of all inverses is the set

(11) $$X^{-1} = \{x^{-1} : x \in X\}.$$

The members of the union $Z = X \cup X^{-1}$ are defined to be distinct from one another. So $|X| = |X^{-1}|$ and $|Z| = |X \cup X^{-1}| = |X| + |X^{-1}|$.

Next, define a *word* in X to be any finite string of elements of $Z = X \cup X^{-1}$. The *empty word e* is the word with no symbols at all. If x appears immediately next to its inverse x^{-1} (in either order), the word may be simplified by omitting the $x x^{-1}$ or $x^{-1} x$ pair. If an element x of X appears consecutively k times in a word, the substring may be abbreviated as the power x^k; likewise, k consecutive x^{-1}s is abbreviated x^{-k}. For example, if $X = \{x_1, x_2\}$, then $X^{-1} = \{x_1^{-1}, x_2^{-1}\}$, and the word $x_1 x_1 x_1 x_1 x_1^{-1} x_2^{-1} x_2^{-1} x_1^{-1} x_1$ may be simplified to $x_1^3 x_2^{-2}$. A word that cannot be simplified further is called *reduced*. A reduced word a may alternatively be considered the (representative of the) equivalence class $[a]$ of all words reducible to a.

The *free group* GX is defined to be the group of all reduced words in X, with concatenation of words (followed by reduction if necessary) as group operation. The identity is the empty word e. Analogously to free monoids, members of X are called *generators*, and if X is a finite set, X^* is called *finitely generated.* The free group GX over the generator set X is a group in which each element can be *uniquely* described as a finite product of powers, of the form

$$a = x_1^{n_1} x_2^{n_2} \cdots x_k^{n_k}, \tag{12}$$

where each $x_i \in X$, adjacent x_is are distinct, and n_is are non-zero integers (but one may have $k = 0$, in which case $a = e$). The inverse of a is the word $a^{-1} = x_k^{-n_k} \cdots x_2^{-n_2} x_1^{-n_1}$.

A free group is 'free' in the sense, again, that it is subject to no *nontrivial* relations. That it is a group (as opposed to monoid) means there are no relations among the words other than those required by the three group axioms ($g1$) associativity, ($g2$) identity, and ($g3$) inverses; i.e., equations (3)–(5). Note that in the reduction schemes introduced above, only the removal of xx^{-1} and $x^{-1}x$ from a word is a consequence of a group axiom [which in this case is ($g3$)]. The simplification of $xx\cdots x$ to the power x^k is just an abbreviation; so $xx\cdots x = x^k$ is not a nontrivial relation, $xx\cdots x$ and x^k being alternate representations of the same element in GX.

A free group (with a non-empty set of generators) is necessarily infinite. This is because in a finite group G of order n, each element $a \in G$ satisfies the relation $a^n = e$, which is a constricting relation on G. If the generator is a singleton set $X = \{x\}$, then the free group GX over X is the infinite cyclic group $GX = \{x^n : n \in \mathbb{Z}\} \cong \langle \mathbb{Z}, + \rangle$. When X contains more then one element, the free group GX is non-abelian (since commutativity would require the nontrivial relation $x_1 x_2 = x_2 x_1$ for $x_1, x_2 \in X$).

12.12 Universal Group The free group GX is the *universal group* generated by the set X, as formalized by the following universal property. Given any mapping g from X to (the underlying set of) a group H, there exists a unique homomorphism $f : GX \to H$ that makes the following diagram commute:

(13) 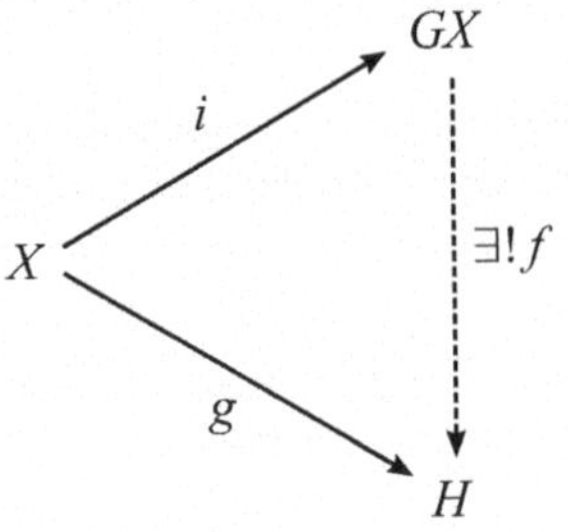

where $i : X \to GX$ is the inclusion map from X into (the underlying set of) GX.

As in the case of free monoids, one may construct f explicitly. First, f must send the empty word $e \in GX$ to the identity of H. Consider the inclusion map $i : X \to GX$ as sending each symbol x in X to a word $i(x) = x$ in GX consisting of that symbol. f has to agree with g on the elements of X, so one must define $f(i(x)) = g(x)$. For the remaining words in GX (consisting of strings of more than one symbol), f may be uniquely extended since it is a homomorphism: if $a = x_1^{n_1} x_2^{n_2} \cdots x_k^{n_k}$, then $f(a) = g(x_1)^{n_1} g(x_2)^{n_2} \cdots g(x_k)^{n_k}$. A fortiori, $f(a^{-1}) = f(x_k^{-n_k} \cdots x_2^{-n_2} x_1^{-n_1}) = g(x_k)^{-n_k} \cdots g(x_2)^{-n_2} g(x_1)^{-n_1} = f(a)^{-1}$. Thus one likewise concludes that the action of a (group) homomorpism with a free group GX as domain is uniquely determined by its values on the generator set X.

One notes that the commutative diagrams (10) and (13) are exactly analogous. Being characterized by a universal property is in fact the standard feature of free objects in universal algebra. The construction of the free objects is, indeed, a functor from the category of sets to the category of algebraic objects under consideration.

Free–Forgetful Adjunction

12.13 Universal and Free Recall (Definition 0.16) that a *concrete category* **C** is a category equipped with a faithful functor $F : \mathbf{C} \to \mathbf{Set}$. The faithfulness of F allows the consideration of **C**-objects as *sets with additional structure*, and of **C**-morphisms as *structure-preserving mappings*. The functor $F : \mathbf{C} \to \mathbf{Set}$ then, in essence, 'forgets' the additional structure of the objects and hence the structure-preserving aspect of the mappings; it is therefore called the *forgetful functor*.

Let $X \in \mathbb{O}\mathbf{Set}$ be a set and $A \in \mathbb{O}\mathbf{C}$ a $\mathbf{C}$-object. Let $i \in \mathbf{Set}(X, FA)$, i.e. the mapping $i : X \to FA$. One says that A is the *free object* on X (with respect to i) if and only if this universal property is satisfied: for any $\mathbf{C}$-object $B \in \mathbb{O}\mathbf{C}$ and any mapping $g : X \to FB$, there exists a unique $\mathbf{C}$-morphism $f \in \mathbf{C}(A, B)$ such that $g = Ff \circ i$. That is, the following diagram commutes:

(14)

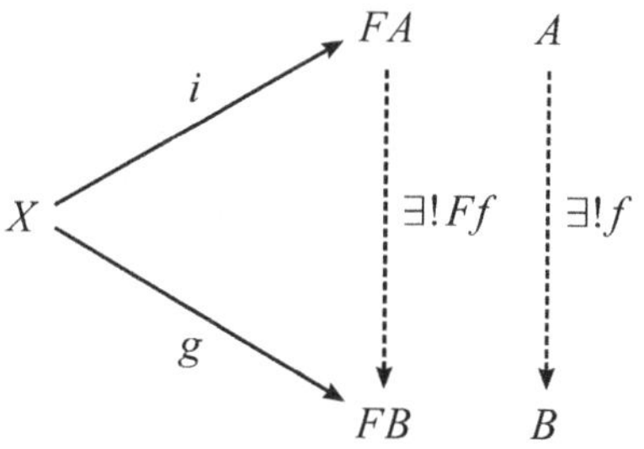

It is through this universal property that the free functor $G : \mathbf{Set} \to \mathbf{C}$ defining the free object A from the set X (called its set of *generators*, or its *basis*) becomes left adjoint to the forgetful functor $F : \mathbf{C} \to \mathbf{Set}$. The free $\mathbf{C}$-object over a set X may be considered as a generic algebraic structure over X: the only equations that hold between elements of the free object are those that follow from the defining axioms of the algebraic structure of $\mathbf{C}$-objects.

12.14 The Set–Vct Adjunction The canonical example of this free–forgetful adjunction is from the category **Set** to the category **Vct** (of vector spaces and linear transformations):

$$\mathbf{Vct} \underset{F}{\overset{G}{\rightleftarrows}} \mathbf{Set}. \tag{15}$$

The right adjoint is the forgetful functor $F : \mathbf{Vct} \to \mathbf{Set}$ that assigns to a vector space V the set FV of all its vectors, and its left adjoint is $G : \mathbf{Set} \to \mathbf{Vct}$ that sends a set X to the vector space GX with basis X (which is the free **Vct**-object over X, those equivalent formal combinations of all members of X modulo the vector space axioms; i.e., all formal linear combinations of members of X). The bijections $\varphi_{V,X} : \mathbf{Vct}(GX, V) \to \mathbf{Set}(X, FV)$ are $\varphi_{V,X} : T \mapsto T|_X$ that send a linear transformation $T : GX \to V$ to its restriction mapping on the basis X.

Note that the inverse $\varphi_{V,X}^{-1}$ extends a mapping defined on a basis X to the whole vector space GX, which is the linear-algebraic theorem that a linear transformation is uniquely determined by its values on a basis. The cardinality of the basis X, i.e., the rank of the vector space GX, is more commonly called its *dimension*. Generators of free objects are, indeed, the direct generalization to categories of the notion of basis of a vector space.

12.15 The Set–Grp Adjunction Another example of the free–forgetful adjunction is from the category **Set** to the concrete category **Grp** (of groups and homomorphisms; cf. Section 0.16 and *ML*: A.6, A.26, A.36):

$$(16) \qquad \mathbf{Grp} \underset{F}{\overset{G}{\rightleftarrows}} \mathbf{Set}.$$

(This expands the **Set–Vct** adjunction, if one notes that a vector space is an abelian group under its addition operation.)

The right adjoint is the *forgetful functor* $F : \mathbf{Grp} \to \mathbf{Set}$ that sends a group to its underlying set and a homomorphism to its underlying mapping (i.e., 'forgetting' the group structure).

The left adjoint is the *free functor* $G : \mathbf{Set} \to \mathbf{Grp}$ that assigns to each set X the free group GX generated by (the elements of) X. A homomorphism with domain GX is fully determined by its action on the elements of X that are generators of GX, thus a mapping $g \in \mathbf{Set}(X, Y)$ naturally extends to a homomorphism $Gg \in \mathbf{Grp}(GX, GY)$.

Indeed, the correspondence between homomorphisms from the free group GX to a group A and mappings from the set X to the set FA precisely defines the hom-set adjunction

$$(17) \qquad \varphi_{A,X} : \mathbf{Grp}(GX, A) \to \mathbf{Set}(X, FA).$$

12.16 Terminal and Initial Morphisms For each group A, GFA is the free group generated by FA, the underlying set of A. Let $\bar{\eta}_A : GFA \to A$ be the group homomorphism which sends the generators of GFA to the elements of A, then each pair $(FA,\ \bar{\eta}_A)$ is a terminal morphism from G to A in **Grp**. Dually, for each set X, the set FGX is the underlying set of the free group GX generated by X. Let $\eta_X : X \to FGX$ be the inclusion map (Definitions 1.2 and 3.6), then each pair $(GX,\ \eta_X)$ is an initial morphism from X to F in **Set**. The counit–unit equations

$$F\,\bar{\eta} \circ \eta_F = 1_F \quad (\text{in } \mathbf{Set}^{\mathbf{Grp}}) \qquad (18)$$
$$\bar{\eta}_G \circ G\eta = 1_G \quad (\text{in } \mathbf{Grp}^{\mathbf{Set}}) \cdot$$

may be verified through the equalities

$$F\,\bar{\eta}_A \circ \eta_{FA} = 1_{FA} \in \mathbf{Set}(FA, FA) \qquad (A \in \mathbb{O}\mathbf{Grp}) \qquad (19)$$
$$\bar{\eta}_{GX} \circ G\,\eta_X = 1_{GX} \in \mathbf{Grp}(GX, GX) \quad (X \in \mathbb{O}\mathbf{Set}) \;,$$

by examining the actions of the **Set** and **Grp** morphisms involved.

12.17 Forgetfulness Galore The adjunction from **Set** to **Grp** is but one specific example of the very common

$$\text{free} \dashv \text{forgetful} \qquad (20)$$

adjunction in mathematics. Free objects are all examples of a left adjoint to a forgetful functor that assigns to an algebraic object its underlying set. Forgetful functors tend to have left adjoints (but not so often right adjoints). See Mac Lane [1997, Chapter IV, Section 2], for a list of examples of left adjoints of typical forgetful functors.

12.18 When a Proset Forgets The forgetful functor $F : \mathbf{Pro} \to \mathbf{Set}$ has both a left adjoint and a right adjoint. Its left adjoint is $\Delta : \mathbf{Set} \to \mathbf{Pro}$ that sends a set X to (X, Δ_X), with the diagonal (identity) relation as preorder. Its right adjoint is $U : \mathbf{Set} \to \mathbf{Pro}$ that sends a set X to $(X,\ U)$, with the universal relation as preorder.

$$\mathbf{Pro} \ \underset{F}{\overset{\Delta}{\leftrightarrows}}\ \mathbf{Set}; \quad \mathbf{Set} \ \underset{U}{\overset{F}{\rightleftarrows}}\ \mathbf{Pro}. \qquad (21)$$

A mapping is trivially isotone if it only has to preserve the identity relation, or if all elements in its codomain are related.

12.19 Shades of Forgetfulness The term ‘forgetful functor’ does not have to apply only to a functor that forgets the algebra; it may simply ‘forget’ some or all of the structures, properties, objects, or morphisms of its domain category. For example, the forgetful functor $F : \mathbf{Grp} \to \mathbf{Set}$ forgets the structure of a group and the property of admitting a group structure. Its left adjoint the free group functor $G : \mathbf{Set} \to \mathbf{Grp}$ is, however, also forgetful: G is not injective on objects, and so forgets them, in the sense (just like a non-injective mapping on any hierarchical level; cf. Section 9.3) that distinct **Set**-objects are no longer distinguishable in the subcategory $G(\mathbf{Set})$ of **Grp**. G, in short, forgets the structure of a set of free generators and the property of 'being free'.

If a functor is not injective on objects, then it forgets objects. If a functor is not faithful, then it forgets morphisms. A functor that is essentially surjective on objects and faithful may forget structures. A functor, even if faithful and full, may still forget properties. Every functor forgets, to various levels and degrees. What entities and how much a functor forgets may, indeed, be used in its classification.

The left adjoint, in its construction of "free structures", are the means to the end of building something from nothing (*ex nihilo aliquid*, as it were). The idea that a mapping is completely determined by its actions on a few generator elements (basis) is a way of rebuilding something that has been forgotten. Using adjoints, one may build back causal structures, i.e., entailment networks, or 'abstract block diagrams' in any category. One sees an immediate realization in (M, R)-systems in replication, which may be viewed as building back repair processes from their products alone; recall (Section 8.17) the formal causes of the three kinds of replication $\mathrm{M}_1^{\leftarrow} : (b, f) \mapsto \Phi$, $\mathrm{M}_2^{\leftarrow} : (\boldsymbol{b}, b) \mapsto \Phi$, and $\mathrm{M}_3^{\leftarrow} : (a, a) \mapsto \Phi$, in which the material causes are final causes of existent efficient causes. In molecular terms, replication is how proteins control DNA by 'storing' the replication instructions, just as DNA stores (i.e., serves as a 'memory device' for) the repair instructions. Incidentally, mutation and other genetic modifications imply that DNA is RAM, not ROM! Proteins remain the 'workhorses' of the cell: metabolism is continual; repair and replication are occasional.

In category theory one often has two categories that are not on the same conceptual level. This is most clear in the case of free–forgetful adjunctions. Each adjunction provides a dictionary between two categories that are not necessarily on an equal footing.

What an abstraction forgets is *scaffolding*; what it retains is *site*.

13
Power and Riches

— Georg Friedrich Händel
[George Frederic Handel] (1685–1759)
Messiah (HWV 56)
Worthy is the Lamb

A. H. Louie, *Intangible Life*, Anticipation Science 2,
DOI 10.1007/978-3-319-65409-6_14

In *IL*: Part I, our exploration of the power set functor culminated in the functorial commutative diagram

(1)

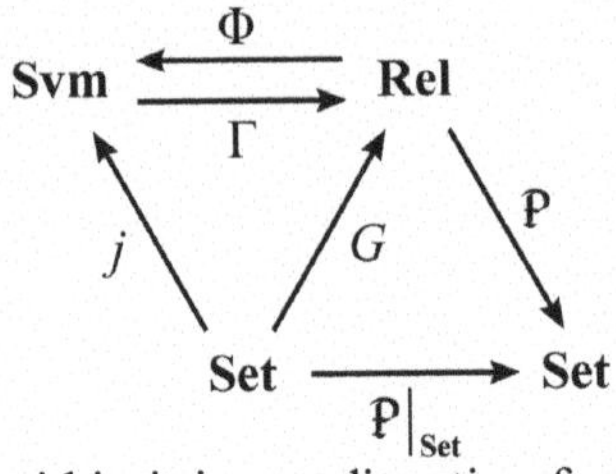

(cf. Section 4.1). Hidden within it is an adjunction from **Set** to **Rel**,

(2) $$\mathbf{Rel} \underset{\mathrm{P}}{\overset{G}{\leftrightarrows}} \mathbf{Set}.$$

(Note that while arrow diagram (1) is commutative, arrow diagram (2) is not, the latter only serving to indicate the dual domains and codomains of the functors.) The consequences of the power set functor P as adjoint are a cornucopia of riches.

Graph and Power Set Functors

13.1 The Left Adjoint Recall (Section 2.5) the *graph functor* $G : \mathbf{Set} \to \mathbf{Rel}$ defined by

(3) $$G : \begin{cases} X \mapsto X & (X \in \mathfrak{O}\mathbf{Set}) \\ [f : X \to Y] \mapsto [f \subset X \times Y] & (f \in \mathfrak{A}\mathbf{Set}) \end{cases},$$

sending a set to itself and a mapping $f : X \to Y$ to its graph

(4) $$Gf = \{(x, f(x)) : x \in X\} \subset X \times Y.$$

Theorem 4.2.iii is

(5) $$G = \Gamma \circ j : \mathbf{Set} \to \mathbf{Svm} \to \mathbf{Rel},$$

which is the trace of $[f : X \to Y] \mapsto [f \subset X \times Y]$ factored through the mapping f's set-valued mapping counterpart $\left[\Gamma_f : x \mapsto \{f(x)\}\right]$.

13.2 The Right Adjoint The right adjoint of the graph functor G is the *power set functor* (Section 3.15) $\mathrm{P} : \mathbf{Rel} \to \mathbf{Set}$ defined by

(6) $$\mathrm{P} : \begin{cases} A \mapsto \mathrm{P}A & (A \in \mathfrak{O}\,\mathbf{Rel}) \\ [R \subset A \times B] \mapsto \left[\mathrm{P}R : E \mapsto \operatorname{ran}\left(R|_E\right)\right] & (R \in \mathfrak{A}\,\mathbf{Rel}) \end{cases}.$$

That is, P takes a set to its power set, and takes a relation $R \subset A \times B$ to the mapping $\mathrm{P}R : \mathrm{P}A \to \mathrm{P}B$ that sends $E \subset A$ to the range of the restriction $R|_E$ of R to E (which is the set of all right R-relatives of elements of E; Section 1.26):

(7) $$(\mathrm{P}R)(E) = \operatorname{ran}\left(R|_E\right) = \{b \in B : \exists a \in E(a, b) \in R\} \subset B.$$

Equivalently,

(8) $$\mathrm{P}R(E) = \bigcup_{a \in E} \mathrm{P}R(\{a\}) \subset B,$$

so $\mathrm{P}R : \mathrm{P}A \to \mathrm{P}B$ sends $E \subset A$ to its *image* under the set-valued mapping $R : A \multimap B$ (cf. Section 3.19).

Theorem 4.2.v is

(9) $$\mathrm{P}|_{\mathbf{Set}} = \mathrm{P} \circ G : \mathbf{Set} \to \mathbf{Rel} \to \mathbf{Set},$$

which says that the standard power set functor $\mathrm{P}|_{\mathbf{Set}} : \mathbf{Set} \to \mathbf{Set}$ takes $A \mapsto \mathrm{P}A$ and $f \subset A \times B$ to the power set mapping $\mathrm{P}f : \mathrm{P}A \to \mathrm{P}B$ that sends $E \subset A$ to its image $f(E) \subset B$:

(10) $$\mathrm{P}|_{\mathbf{Set}} : \begin{cases} A \mapsto \mathrm{P}A & (A \in \mathfrak{O}\,\mathbf{Set}) \\ [f : A \to B] \mapsto [\mathrm{P}f : E \mapsto f(E)] & (f \in \mathfrak{A}\,\mathbf{Set}) \end{cases}.$$

13.3 Bijection For $A \in \mathbf{Rel}$ and $X \in \mathbf{Set}$, the adjunction bijection

(11) $$\varphi_{A,X} : \mathbf{Rel}(GX, A) \to \mathbf{Set}(X, \mathrm{P}A)$$

sends $R \subset GX \times A = X \times A$ to the mapping $\varphi_{A,X}(R) : X \to \mathrm{P}A$ defined by, for $x \in X$,

(12) $$\varphi_{A,X}(R) : x \mapsto \{a \in A : (x, a) \in R\}.$$

The mapping $\varphi_{A,X}(R) : X \to \mathrm{P}A$ is, of course, by Definition 3.1 also a set-valued mapping $\varphi_{A,X}(R) : X \multimap A$. It is in fact the set-valued mapping $\Phi_R : X \multimap A$ that appeared in Section 3.13, with the value at $x \in X$ the set of all right R-relatives of x (or of the singleton set $\{x\}$),

(13) $$\Phi_R(x) = \operatorname{ran}\left(R|_{\{x\}}\right).$$

Recall (Section 3.14) that the assignment $R \mapsto \Phi_R$ is the arrow map of the inverse graph functor $\Phi : \mathbf{Rel} \to \mathbf{Svm}$ that defines the isomorphism of categories

(14) $$\mathbf{Svm} \underset{\Gamma}{\overset{\Phi}{\leftrightarrows}} \mathbf{Rel}$$

(Diagram (14) is commutative, the commutativity being the statements $\Gamma \circ \Phi = I_{\mathbf{Rel}}$ and $\Phi \circ \Gamma = I_{\mathbf{Svm}}$, the very definition of the $\mathbf{Rel} \cong \mathbf{Svm}$ isomorphism.)

13.4 Inverse The inverse adjunction bijection is

(15) $$\varphi_{A,X}^{-1} : \mathbf{Set}(X, \mathrm{P}A) \to \mathbf{Rel}(GX, A).$$

It sends $f : X \to \mathrm{P}A$ to the relation $\varphi_{A,X}^{-1}(f) \subset GX \times A = X \times A$ defined by

(16) $$\varphi_{A,X}^{-1}(f) = \{(x, a) \in X \times A : a \in f(x) \subset A\}.$$

Likewise, the relation $\varphi_{A,X}^{-1}(f) \subset X \times A$ corresponds to an equivalent set-valued mapping $\varphi_{A,X}^{-1}(f) : X \multimap A$.

Identity and Converse Membership

13.5 Unit and Identity Map The natural transformation

$$\eta : I_{\mathbf{Set}} \to \mathrm{P} \circ G \tag{17}$$

maps between functors from **Set** to **Set**, i.e.,

$$\mathbf{Set} \underset{\mathrm{P}\circ G}{\overset{I_{\mathbf{Set}}}{\rightrightarrows}} \mathbf{Set} \tag{18}$$

That the unit η is a natural transformation (Definition 0.18) means:

i. for each **Set**-object (set) X, there is a **Set**-morphism (mapping) $\eta_X \in \mathbf{Set}(I_{\mathbf{Set}}X, (\mathrm{P} \circ G)X)$.

Now $(\mathrm{P} \circ G)X = \mathrm{P}X$, so $\eta_X \in \mathbf{Set}(X, \mathrm{P}X)$.

The components $\eta_X \in \mathbf{Set}(X, \mathrm{P}X)$ of the natural transformation η have the property

($t1$) for each **Set**-morphism (mapping) $f \in \mathbf{Set}(X, Y)$,

$$(\mathrm{P} \circ G)f \circ \eta_X = \eta_Y \circ I_{\mathbf{Set}}f. \tag{19}$$

If $f \in \mathbf{Set}(X, Y)$, then $Gf = \{(x, f(x)) : x \in X\} \subset X \times Y$; whence $(\mathrm{P} \circ G)f : \mathrm{P}X \to \mathrm{P}Y$, defined by, for $E \subset X$,

$$(\mathrm{P} \circ G)f : E \mapsto \{y \in Y : y = f(x) \text{ for some } x \in E\} = f(E) \subset Y; \tag{20}$$

i.e., $(\mathrm{P} \circ G)f$ is the power set mapping $\mathrm{P}f : \mathrm{P}X \to \mathrm{P}Y$.

One notes that $(\mathrm{P} \circ G)X = \mathrm{P}X$ and $(\mathrm{P} \circ G)f = \mathrm{P}f$, so in particular $\mathrm{P} \circ G \neq I_{\mathbf{Set}}$ (thus P and G are not inverses of each other). Indeed, $\mathrm{P} \circ G : \mathbf{Set} \to \mathbf{Set}$ is the covariant power set functor on **Set** (Definition 3.18):

$$\mathrm{P} \circ G = \mathrm{P}|_{\mathbf{Set}} : \begin{cases} X \mapsto \mathrm{P}X & (X \in \mathfrak{O}\,\mathbf{Set}) \\ [f : X \to Y] \mapsto [\mathrm{P}f : E \mapsto f(E)] & (f \in \mathfrak{A}\,\mathbf{Set}) \end{cases}. \tag{21}$$

Evidently, $I_{\mathbf{Set}}f = f$. So equation (19) simplifies to

(22) $$\mathrm{P}f \circ \eta_X = \eta_Y \circ f.$$

One naturally defines $\eta_X \in \mathbf{Set}(X, \mathrm{P}X)$ as the identity set-valued mapping (Definition 3.6) $1_X : X \mathrel{-\!\sqsubset} X$. In other words, the component of the unit η at set X is

(23) $$\eta_X = 1_X : x \mapsto \{x\}.$$

One then readily verifies that both sides of the equation (22) represent the single-valued set-valued mapping $j_f : X \mathrel{-\!\sqsubset} Y$ (Definition 3.2),

(24) $$j_f : x \mapsto \{f(x)\}.$$

The two sides of the commutativity equation (19) [and therefore equation (22)] are compositions that map sequentially thus:

(25) $$\begin{cases} (\mathrm{P} \circ G)f \circ \eta_X = \mathrm{P}f \circ \eta_X : X \to \mathrm{P}X \to \mathrm{P}Y, & x \mapsto \{x\} \mapsto f(\{x\}) = \{f(x)\} \\ \eta_Y \circ I_{\mathbf{Set}}f = \eta_Y \circ f : X \to Y \to \mathrm{P}Y, & x \mapsto f(x) \mapsto \{f(x)\} \end{cases}.$$

(26)

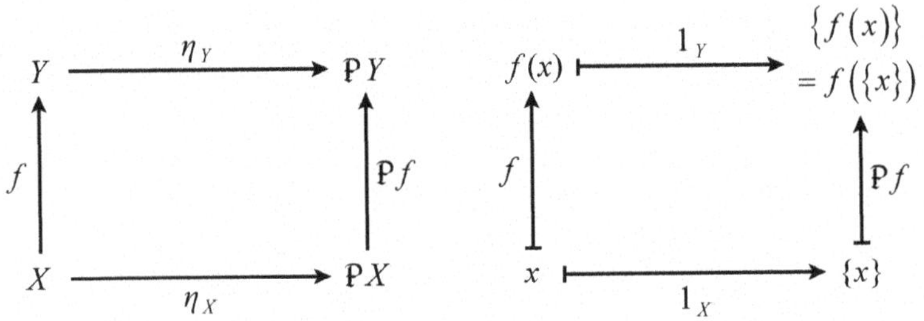

In sum, the unit $\eta : I_{\mathbf{Set}} \to \mathrm{P} \circ G$ is the natural transformation from the identity functor $I_{\mathbf{Set}}$ to the (standard) power set functor $\mathrm{P} : \mathbf{Set} \to \mathbf{Set}$.

13.6 Counit and Converse Membership The natural transformation

(27) $$\bar{\eta} : G \circ \mathrm{P} \to I_{\mathbf{Rel}}$$

maps between functors from **Rel** to **Rel**, i.e.,

(28) $$\mathbf{Rel} \underset{I_{\mathbf{Set}}}{\overset{G \circ \text{Ᵽ}}{\rightrightarrows}} \mathbf{Rel}$$

That the counit $\overline{\eta}$ is a natural transformation means:
i. for each **Rel**-object A, there is a **Rel**-morphism $\overline{\eta}_A \in \mathbf{Rel}((G \circ \text{Ᵽ})A, I_{\mathbf{Rel}}A)$.

Now $(G \circ \text{Ᵽ})A = G(\text{Ᵽ}A) = \text{Ᵽ}A$, so $\overline{\eta}_A \in \mathbf{Rel}(\text{Ᵽ}A, A) = \text{Ᵽ}(\text{Ᵽ}A \times A)$. These components $\overline{\eta}_A$ of the natural transformation $\overline{\eta}$ satisfy the property that
($t1$) for each **Rel**-morphism $R \in \mathbf{Rel}(A, B)$,

(29) $$I_{\mathbf{Rel}}R \circ \overline{\eta}_A = \overline{\eta}_B \circ (G \circ \text{Ᵽ})R.$$

Each of the four morphisms involved in the commutativity equation are **Rel**-morphisms, i.e., relations/set-valued mappings.

If $R \in \mathbf{Rel}(A, B) = \text{Ᵽ}(A \times B)$, i.e., $R \subset A \times B$, then $\text{Ᵽ}R : \text{Ᵽ}A \to \text{Ᵽ}B$ is the (single-valued) power set mapping defined, for $E \subset A$, by

(30) $$\text{Ᵽ}R : E \mapsto \{b \in B : (a, b) \in R \text{ for some } a \in E\} = \text{Ᵽ}R(E) \in \text{Ᵽ}B,$$

which is the set of all right R-relatives of elements of E (cf. Section 3.13),

(31) $$\operatorname{ran}\left(R|_E\right) = \text{Ᵽ}R(E) \subset B.$$

Thus

(32) $$(G \circ \text{Ᵽ})R = G(\text{Ᵽ}R) = \{(E, \text{Ᵽ}R(E)) : E \subset A\} \in \mathbf{Rel}(\text{Ᵽ}A, \text{Ᵽ}B),$$

which is the graph of $\text{Ᵽ}R$ (cf. Section 3.20). One may note that $(G \circ \text{Ᵽ})A = \text{Ᵽ}A$ and $(G \circ \text{Ᵽ})R =$ the graph of $\text{Ᵽ}R$, so again $G \circ \text{Ᵽ} \neq I_{\mathbf{Rel}}$.

Evidently, $I_{\mathbf{Rel}}R = R$. So equation (29) simplifies to

(33) $$R \circ \overline{\eta}_A = \overline{\eta}_B \circ G(\text{Ᵽ}R).$$

When the four morphisms involved are interpreted as set-valued mappings, viz.

(34)
$$R : A \multimap B$$
$$\overline{\eta}_A : \text{P}A \multimap A$$
$$\overline{\eta}_B : \text{P}B \multimap B$$
$$G(\text{P}R) : \text{P}A \multimap \text{P}B$$

the commutativity equation (33) is that of their sequential compositions.

Define $\overline{\eta}_A$ as the converse membership relation (Definition 1.23)

(35) $$\overline{\eta}_A = \ni_A = \{(E, a) : E \subset A, a \in E\} \subset \text{P}A \times A.$$

It is therefore the set-valued mapping

(36) $$\overline{\eta}_A = \ni_A : \text{P}A \multimap A$$

that sends a subset $E \subset A$ to the set of all elements it contains, i.e., E itself; whence

(37) $$\ni_A : E \mapsto E.$$

In the set-valued mapping's equivalent Definition 3.1 as the single-valued mapping

(38) $$\ni_A : \text{P}A \to \text{P}A,$$

one therefore has, in fact,

(39) $$\overline{\eta}_A = \ni_A = 1_{\text{P}A} \in \mathbf{Set}(\text{P}A, \text{P}A).$$

With this 'converse membership' definition of the counit components $\overline{\eta}_A$, one may proceed to verify equation (33) [and therefore equation (29)]. The left-hand side is the sequential composition

(40) $$R \circ \overline{\eta}_A : \text{P}A \multimap A \multimap B;$$

it sends a set $E \in \text{P}A$ to $E \subset A$ and then to $\text{ran}(R|_E) \subset B$, the set of all right R-relatives of elements of E:

(41) $$E \mapsto E \mapsto \mathrm{ran}(\mathrm{R}|_E).$$

The right-hand side is the sequential composition

(42) $$\overline{\eta}_B \circ G(\mathcal{P}R) : \mathcal{P}A \multimap \mathcal{P}B \multimap B;$$

it sends a set $E \in \mathcal{P}A$ to $\mathcal{P}R(E) \in \mathcal{P}B$ and then to $\mathcal{P}R(E) \subset B$:

(43) $$E \mapsto \mathcal{P}R(E) \mapsto \mathcal{P}R(E).$$

The equality in (33) is then established by noting that $\mathrm{ran}(R|_E) = \mathcal{P}R(E)$ [as in (31) above], both representing the set of all right R-relatives of elements of E.

(44)

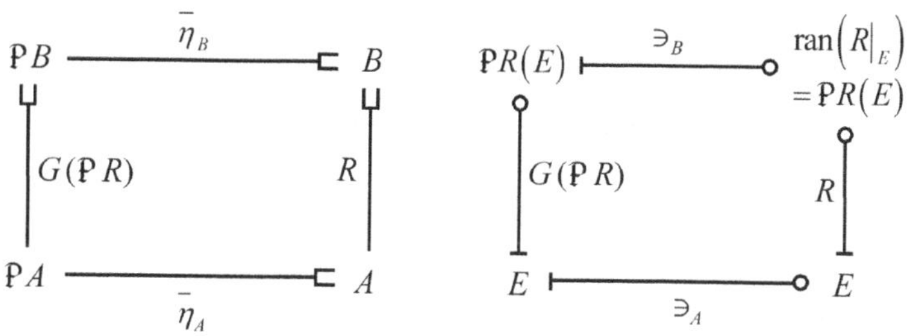

Note that in the element-chase commutative diagram on the right, I have used the 'hollow-circle-headed arrow' (introduced in Section 6.4) notation for set-valued mappings.

13.7 Triangles The counit–unit equations

(45) $$\begin{aligned} &\mathcal{P}\overline{\eta} \circ \eta_{\mathcal{P}} = 1_{\mathcal{P}} \quad (\text{in } \mathbf{Set}^{\mathbf{Rel}}) \\ &\overline{\eta}_G \circ G\eta = 1_G \quad (\text{in } \mathbf{Rel}^{\mathbf{Set}}) \end{aligned}.$$

may be verified through the equalities

(46) $$\begin{aligned} &\mathcal{P}\overline{\eta}_A \circ \eta_{\mathcal{P}A} = 1_{\mathcal{P}A} \in \mathbf{set}(\mathcal{P}A, \mathcal{P}A) \qquad (A \in \mathbf{ORel}) \\ &\overline{\eta}_{GX} \circ G\eta_X = 1_{GX} \in \mathbf{Rel}(\mathcal{P}X, \mathcal{P}X) \quad (X \in \mathbf{Oset}) \end{aligned},$$

by examining the actions of the **Set** and **Rel** morphisms involved.

Explicitly, let $E \in \mathrm{P}A$; then

(47) $$(\mathrm{P}\overline{\eta}_A \circ \eta_{\mathrm{P}A})(E) = \mathrm{P}\overline{\eta}_A(\eta_A(E)) = \mathrm{P}\overline{\eta}_A(\{E\}) = E = 1_{\mathrm{P}A}(E).$$

If $x \in GX = X$, then

(48) $$(\overline{\eta}_{GX} \circ G\eta_{GX})(x) = \overline{\eta}_{GX}(G\eta_X(x)) = \overline{\eta}_{GX}(\{x\}) = \{x\} = 1_{GX}(x).$$

(49)

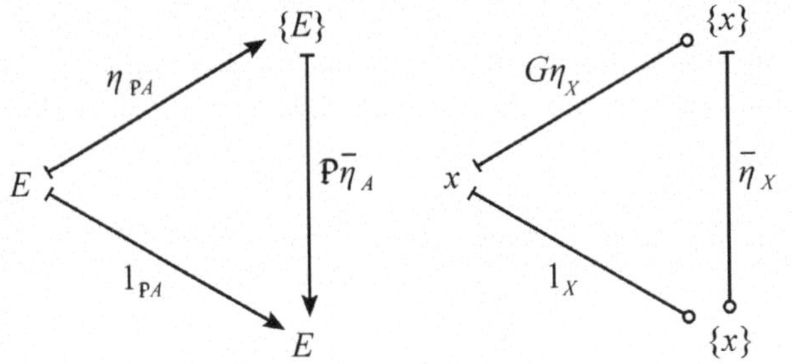

13.8 Universal Morphisms For each set X, there exists an initial morphism from X to P. Explicitly, the set $GX = X$ and the mapping $\eta_X = 1_X : X \to \mathrm{P}X$ are such that, for any mapping $g : X \to \mathrm{P}A$, there is a unique **Rel**-morphism $f : X \dashv\sqsubset A$ that satisfies $\mathrm{P}f \circ \eta_X = g$. Namely, one defines, for $x \in X$,

(50) $$f(x) = g(x) \in \mathrm{P}A.$$

Then

(51) $$\mathrm{P}f \circ \eta_X(x) = \mathrm{P}f(\eta_X(x)) = \mathrm{P}f(\{x\}) = f(x) = g(x).$$

(52)

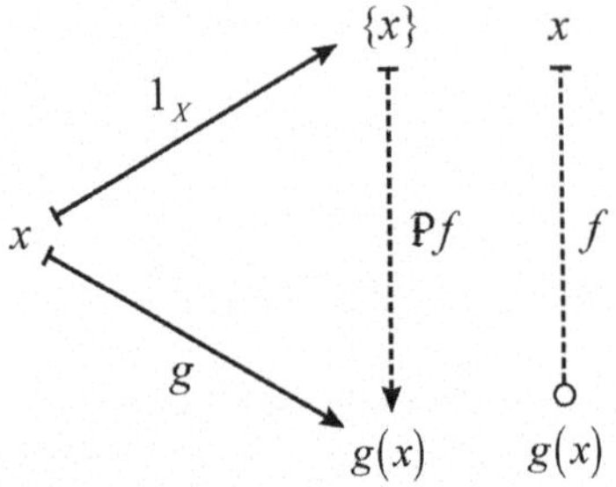

Dually, for each set A, there exists a terminal morphism from G to A. The power set $\mathrm{P}A$ and the **Rel**-morphism $\overline{\eta}_A : \mathrm{P}A \to A$ are such that for any **Rel**-morphism $f : X \mathrel{-\!\!\sqsubset} A$, there is a unique mapping $g : X \to \mathrm{P}A$ satisfying $\overline{\eta}_A \circ Gg = f$. Namely, one defines, for $x \in X$,

$$g(x) = f(x) \in \mathrm{P}A. \tag{53}$$

Then

$$\overline{\eta}_A \circ Gg(x) = \overline{\eta}_A(Gg(x)) = \overline{\eta}_A(g(\{x\})) = g(x) = f(x). \tag{54}$$

(55)

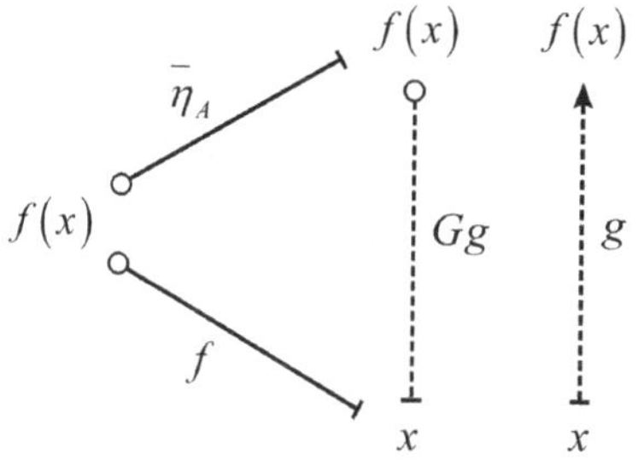

Adjacency Matrices

The unit commutativity equations (19) and (22), $\mathrm{P}f \circ \eta_X = \eta_Y \circ f$, are relatively straightforwardly resolved into the mapping (24), $j_f : X \to \mathrm{P}Y$ for both sides. The counit commutativity equations (29) and (33), $R \circ \overline{\eta}_A = \overline{\eta}_B \circ G(\mathrm{P}R)$, on the other hand, is less immediately understood. An illustrative example may illuminate.

13.9 Power Set Mapping and Its Graph Let $A = \{1, 2\}$ and $B = \{a, b, c\}$; their power sets are $\mathrm{P}A = \{\varnothing, \{1\}, \{2\}, \{1, 2\}\}$ and $\mathrm{P}B = \{\varnothing, \{a\}, \{b\}, \{c\}, \{a, b\}, \{a, c\}, \{b, c\}, \{a, b, c\}\}$.

Consider the relation $R = \{(1, a), (2, a), (2, b)\} \subset A \times B$. Its *adjacency matrix* is

(56)
$$[R] = \begin{array}{c} \\ 1 \\ 2 \end{array}\!\!\begin{array}{c} \begin{array}{ccc} a & b & c \end{array} \\ \begin{bmatrix} 1 & 0 & 0 \\ 1 & 1 & 0 \end{bmatrix} \end{array}.$$

(The adjacency matrix $[R]$ (*RL*: 5.2) of a relation $R \subset A \times B$ with finite sets $A = \{a_i : i\}$ and $B = \{b_j : j\}$ has entries defined by

(57)
$$R_{ij} = \begin{cases} 1 & \text{if}\,(a_i, b_j) \in R \\ 0 & \text{if}\,(a_i, b_j) \notin R \end{cases}.)$$

The power set mapping $\mathrm{P}R : \mathrm{P}A \to \mathrm{P}B$ is the single-valued mapping with $\varnothing \mapsto \varnothing$, $\{1\} \mapsto \{a\}$, $\{2\} \mapsto \{a,b\}$, $\{1,2\} \mapsto \{a,b\}$. So, $G(\mathrm{P}R) \subset \mathrm{P}A \times \mathrm{P}B$, the graph of $\mathrm{P}R$, is $G(\mathrm{P}R) = \{(\varnothing,\varnothing),(\{1\},\{a\}),(\{2\},\{a,b\}),(\{1,2\},\{a,b\})\} \subset \mathrm{P}A \times \mathrm{P}B$. Their adjacency matrix $[\mathrm{P}R] = [G(\mathrm{P}R)]$ is

(58)
$$[G(\mathrm{P}R)] = \begin{array}{c} \\ \varnothing \\ \{1\} \\ \{2\} \\ \{1,2\} \end{array}\!\!\begin{array}{c} \begin{array}{cccccccc} \varnothing & \{a\} & \{b\} & \{c\} & \{a,b\} & \{a,c\} & \{b,c\} & \{a,b,c\} \end{array} \\ \begin{bmatrix} 1 & 0 & 0 & 0 & 0 & 0 & 0 & 0 \\ 0 & 1 & 0 & 0 & 0 & 0 & 0 & 0 \\ 0 & 0 & 0 & 0 & 1 & 0 & 0 & 0 \\ 0 & 0 & 0 & 0 & 1 & 0 & 0 & 0 \end{bmatrix} \end{array}$$

Note that $[G(\mathrm{P}R)]$ is the adjacency matrix of a single-valued mapping, so each row contains exactly one '1'.

The single-valued mapping $\mathrm{P}R : \mathrm{P}A \to \mathrm{P}B$ is, by Definition 3.1, also the set-valued mapping $\mathrm{P}R : \mathrm{P}A \multimap B$ (which, likewise, maps $\mathrm{P}R(\varnothing) = \varnothing$, $\mathrm{P}R(\{1\}) = \{a\}$, $\mathrm{P}R(\{2\}) = \{a,b\}$, $\mathrm{P}R(\{1,2\}) = \{a,b\}$), sending a set $E \in \mathrm{P}A$ to $\mathrm{ran}(R|_E) = \mathrm{P}R(E) \subset B$. As the relation $\Gamma_{\mathrm{P}R} \subset \mathrm{P}A \times B$ (its graph; Definition 3.10), it has the adjacency matrix

(59)
$$[\Gamma_{PR}] = \begin{array}{c} \\ \varnothing \\ \{1\} \\ \{2\} \\ \{1,2\} \end{array} \begin{array}{c} \begin{array}{ccc} a & b & c \end{array} \\ \begin{bmatrix} 0 & 0 & 0 \\ 1 & 0 & 0 \\ 1 & 1 & 0 \\ 1 & 1 & 0 \end{bmatrix} \end{array}$$

13.10 Converse Membership Relation The membership relation $\in_X \subset A \times PA$ (Example 1.20) contains the following element–subset ordered-pairs: $(1, \{1\})$, $(2, \{2\})$, $(1, \{1, 2\})$, and $(2, \{1, 2\})$. Thus the converse membership relation of A is

(60) $$\ni_A = \overline{\eta}_A = \{(\{1\}, 1), (\{2\}, 2), (\{1, 2\}, 1), (\{1, 2\}, 2)\} \subset PA \times A.$$

The adjacency matrix of $\in_A$ is

(61)
$$[\in_A] = \begin{array}{c} \\ 1 \\ 2 \end{array} \begin{array}{c} \begin{array}{cccc} \varnothing & \{1\} & \{2\} & \{1,2\} \end{array} \\ \begin{bmatrix} 0 & 1 & 0 & 1 \\ 0 & 0 & 1 & 1 \end{bmatrix} \end{array},$$

and the adjacency matrix of $\ni_A$ is

(62)
$$[\ni_A] = \begin{array}{c} \\ \varnothing \\ \{1\} \\ \{2\} \\ \{1,2\} \end{array} \begin{array}{c} \begin{array}{cc} 1 & 2 \end{array} \\ \begin{bmatrix} 0 & 0 \\ 1 & 0 \\ 0 & 1 \\ 1 & 1 \end{bmatrix} \end{array}$$

Similarly, one has

(63)
$$\begin{aligned} \ni_B &= \overline{\eta}_B \\ &= \left\{ \begin{array}{l} (\{a\}, a), (\{b\}, b), (\{c\}, c), \\ (\{a, b\}, a), (\{a, b\}, b), (\{a, c\}, a), \\ (\{a, c\}, c), (\{b, c\}, b), (\{b, c\}, c), \\ (\{a, b, c\}, a), (\{a, b, c\}, b), (\{a, b, c\}, c) \end{array} \right\}. \\ &\subset PB \times B \end{aligned}$$

and

(64)
$$[\ni_B] = \begin{array}{c} \\ \varnothing \\ \{a\} \\ \{b\} \\ \{c\} \\ \{a,b\} \\ \{a,c\} \\ \{b,c\} \\ \{a,b,c\} \end{array} \begin{array}{c} \begin{array}{ccc} a & b & c \end{array} \\ \begin{bmatrix} 0 & 0 & 0 \\ 1 & 0 & 0 \\ 0 & 1 & 0 \\ 0 & 0 & 1 \\ 1 & 1 & 0 \\ 1 & 0 & 1 \\ 0 & 1 & 1 \\ 1 & 1 & 1 \end{bmatrix} \end{array}$$

13.11 The Counit Commutativity Equation The left-hand side of the counit commutativity equation $R \circ \overline{\eta}_A = \overline{\eta}_B \circ G(\text{Ᵽ}R)$ is the sequential composite of

(65)
$$\overline{\eta}_A = \ni_A \in \mathbf{Rel}(\text{Ᵽ}A, A) = \text{Ᵽ}(\text{Ᵽ}A \times A),$$

followed by

(66)
$$R \in \mathbf{Rel}(A, B) = \text{Ᵽ}(A \times B),$$

i.e.,

(67)
$$R \circ \ni_A : \text{Ᵽ}A -\!\!\sqsubset A -\!\!\sqsubset B,$$

with $\varnothing \mapsto \varnothing \mapsto \varnothing, \{1\} \mapsto 1 \mapsto a$, $\{2\} \mapsto 2 \mapsto \{a,b\}$, $\{1,2\} \mapsto 1 \mapsto a$, $\{1,2\} \mapsto 2 \mapsto \{a,b\}$; thus $R \circ \ni_A : \text{Ᵽ}A -\!\!\sqsubset B$ maps $\varnothing \mapsto \varnothing$, $\{1\} \mapsto a$, $\{2\} \mapsto \{a,b\}$, $\{1,2\} \mapsto \{a,b\}$. In terms of adjacency matrices, the composition is

$$[R \circ \ni_A] = [\ni_A] \circ [R] = \begin{bmatrix} 0 & 0 \\ 1 & 0 \\ 0 & 1 \\ 1 & 1 \end{bmatrix} \circ \begin{bmatrix} 1 & 0 & 0 \\ 1 & 1 & 0 \end{bmatrix}$$

(68)
$$= \begin{array}{c} \\ \varnothing \\ \{1\} \\ \{2\} \\ \{1,2\} \end{array} \begin{array}{c} \begin{array}{ccc} a & b & c \end{array} \\ \begin{bmatrix} 0 & 0 & 0 \\ 1 & 0 & 0 \\ 1 & 1 & 0 \\ 1 & 1 & 0 \end{bmatrix} \end{array} = [\Gamma_{\text{Ᵽ}R}].$$

(For a review of adjacency matrix sequential composition, see *RL*: 5.12.)

The right-hand side of the counit commutativity equation $R \circ \overline{\eta}_A = \overline{\eta}_B \circ G(\text{P}R)$ is the sequential composite of

$$G(\text{P}R) \in \mathbf{Rel}(\text{P}A, \text{P}B) = \text{P}(\text{P}A \times \text{P}B), \tag{69}$$

followed by

$$\overline{\eta}_B = \ni_B \in \mathbf{Rel}(\text{P}B, B) = \text{P}(\text{P}B \times B), \tag{70}$$

i.e.,

$$\ni_B \circ G(\text{P}R) : \text{P}A \multimap \text{P}B \multimap B, \tag{71}$$

with $\varnothing \mapsto \varnothing \mapsto \varnothing$, $\{1\} \mapsto \{a\} \mapsto a$, $\{2\} \mapsto \{a, b\} \mapsto \{a, b\}$, $\{1, 2\} \mapsto \{a, b\} \mapsto \{a, b\}$; thus $\ni_B \circ G(\text{P}R) : \text{P}A \multimap B$ maps $\varnothing \mapsto \varnothing$, $\{1\} \mapsto a$, $\{2\} \mapsto \{a, b\}$, $\{1, 2\} \mapsto \{a, b\}$. In terms of adjacency matrices, the composition is

$$\begin{aligned}
[\ni_B \circ G(\text{P}R)] &= [G(\text{P}R)] \circ [\ni_B] \\
&= \begin{bmatrix} 1 & 0 & 0 & 0 & 0 & 0 & 0 & 0 \\ 0 & 1 & 0 & 0 & 0 & 0 & 0 & 0 \\ 0 & 0 & 0 & 0 & 1 & 0 & 0 & 0 \\ 0 & 0 & 0 & 0 & 1 & 0 & 0 & 0 \end{bmatrix} \circ \begin{bmatrix} 0 & 0 & 0 \\ 1 & 0 & 0 \\ 0 & 1 & 0 \\ 0 & 0 & 1 \\ 1 & 1 & 0 \\ 1 & 0 & 1 \\ 0 & 1 & 1 \\ 1 & 1 & 1 \end{bmatrix} \\
&= \begin{bmatrix} 0 & 0 & 0 \\ 1 & 0 & 0 \\ 1 & 1 & 0 \\ 1 & 1 & 0 \end{bmatrix} = [\Gamma_{\text{P}R}].
\end{aligned} \tag{72}$$

Upon comparing (68) and (72), it is evident that LHS = RHS in $R \circ \overline{\eta}_A = \overline{\eta}_B \circ G(\text{P}R)$.

13.12 Not Inverses We have seen above that this pair $\langle G, \mathfrak{P} \rangle$ of functors,

$$\mathbf{Rel} \underset{\mathfrak{P}}{\overset{G}{\rightleftarrows}} \mathbf{Set}, \tag{73}$$

compose to

$$\mathfrak{P} \circ G = \mathfrak{P}|_{\mathbf{Set}} : \begin{cases} X \mapsto \mathfrak{P}X & (X \in \mathfrak{O}\,\mathbf{Set}) \\ [f : X \to Y] \mapsto [\mathfrak{P}f : E \mapsto f(E)] & (f \in \mathfrak{A}\,\mathbf{Set}) \end{cases} \tag{74}$$

and

$$G \circ \mathfrak{P} : \begin{cases} A \mapsto \mathfrak{P}A & (A \in \mathfrak{O}\mathbf{Rel}) \\ [R \in \mathbf{Rel}(A,B)] & \\ \quad \mapsto [\{(E, \mathfrak{P}R(E)) : E \subset A\} \in \mathbf{Rel}(\mathfrak{P}A, \mathfrak{P}B)] & (R \in \mathfrak{A}\,\mathbf{Rel}) \end{cases}. \tag{75}$$

So evidently $\mathfrak{P} \circ G \neq I_{\mathbf{Set}}$, and $G \circ \mathfrak{P} \neq I_{\mathbf{Rel}}$. Thus the power set functor $\mathfrak{P}$ and the graph functor G are *not* inverses of each other.

13.13 Not Equivalences That the functors $\mathfrak{P}$ and G are *not* equivalences to each other may be seen in a couple of ways.

For each pair of sets A and B, a mapping of hom-sets,

$$\mathfrak{P}_{A,B} : \mathbf{Rel}(A,B) \to \mathbf{Set}(\mathfrak{P}A, \mathfrak{P}B), \tag{76}$$

is defined by the power set functor $\mathfrak{P} : \mathbf{Rel} \to \mathbf{Set}$. (Note that this is not the adjunction bijection (11), $\varphi_{A,X} : \mathbf{Rel}(GX, A) \to \mathbf{Set}(X, \mathfrak{P}A)$.) The mapping $\mathfrak{P}_{A,B}$ is injective, because a relation $R : A \mathrel{-\!\sqsubset} B$ determines uniquely its 'image map' $\mathfrak{P}R : \mathfrak{P}A \to \mathfrak{P}B$. But it is not surjective, because there is no guarantee that an arbitrary mapping $f : \mathfrak{P}A \to \mathfrak{P}B$ is $f = \mathfrak{P}R$ for some relation R. A power set mapping $\mathfrak{P}R$ would have to preserve the 'lattice structure' of $\mathfrak{P}A$ (cf. Properties 3.16). Consider

$$[f] = \begin{array}{c} \\ \varnothing \\ \{1\} \\ \{2\} \\ \{1,2\} \end{array} \begin{array}{c} \begin{array}{cccccccc} \varnothing & \{a\} & \{b\} & \{c\} & \{a,b\} & \{a,c\} & \{b,c\} & \{a,b,c\} \end{array} \\ \begin{bmatrix} 0 & 1 & 0 & 0 & 0 & 0 & 0 & 0 \\ 1 & 0 & 0 & 0 & 0 & 0 & 0 & 0 \\ 0 & 0 & 1 & 0 & 0 & 0 & 0 & 0 \\ 0 & 0 & 0 & 1 & 0 & 0 & 0 & 0 \end{bmatrix} \end{array}, \tag{77}$$

which is a completely legitimate mapping $f : \mathrm{P}A \to \mathrm{P}B$. But $f(\varnothing) \neq \varnothing$, $f(\{1\}) \not\subset f(\{1,2\})$, $f(\{1,2\}) \neq f(\{1\}) \cup f(\{2\})$, etc.; so $f \neq FR$ for any relation $R \subset A \times B$. Thus the functor $\mathrm{P} : \mathbf{Rel} \to \mathbf{Set}$ is faithful but not full. P is also not essentially surjective on objects: there are many sets that are not power sets. For example, a finite set with cardinality that is not a power of 2 cannot be (isomorphic to) a power set (Section 1.14). So, by Theorem 9.18, P is not an equivalence functor.

Likewise, one sees that $G_{X,Y} : \mathbf{Set}(X, Y) \to \mathbf{Rel}(X, Y)$ is injective, because a mapping determines its graph uniquely; but it is not surjective, since there are plenty of relations that are not graphs of mappings. The functor $G : \mathbf{Set} \to \mathbf{Rel}$ is thus also faithful but not full. So, again by Theorem 9.18, G is not an equivalence functor.

Since either of the functors in the pair $\langle G, \mathrm{P}\rangle$ is faithful but not full, one may also invoke Theorem 10.10 to conclude that P and G are not equivalences.

One may also show directly that the unit $\eta : I_{\mathbf{Set}} \to \mathrm{P} \circ G$ and the counit $\overline{\eta} : G \circ \mathrm{P} \to I_{\mathbf{Rel}}$ are natural transformations but not natural isomorphisms. $\eta : I_{\mathbf{Set}} \to \mathrm{P} \circ G$ is a natural isomorphism (in $\mathbf{Set}^{\mathbf{Set}}$) if and only if for each set X, $\eta_X = 1_X : x \mapsto \{x\}$ is an isomorphism (from $I_{\mathbf{Set}}X = X$ to $(\mathrm{P} \circ G)X = \mathrm{P}X$) in $\mathbf{Set}$. But $1_X \in \mathbf{Set}(X, \mathrm{P}X)$ is evidently not an isomorphism; among many reasons, $|X| < |\mathrm{P}X| = 2^{|X|}$. Likewise, $\overline{\eta} : G \circ \mathrm{P} \to I_{\mathbf{Rel}}$ is a natural isomorphism (in $\mathbf{Rel}^{\mathbf{Rel}}$) if and only if for each set A, $\overline{\eta}_A = \ni_A\{(E, a) : a \in E\} \subset \mathrm{P}A \times A$ is an isomorphism (from $(G \circ \mathrm{P})A = \mathrm{P}A$ to $I_{\mathbf{Rel}}A = A$) in $\mathbf{Rel}$. But $\overline{\eta}_A = \ni_A \in \mathbf{Rel}(\mathrm{P}A, A)$ is, for the same cardinality reason, not an isomorphism.

In summary, the $\langle$graph, power set$\rangle$ adjunction $\langle G, \mathrm{P}, \eta, \ni\rangle$ is an example of a true extension of the

(78) $$\text{Inverse} \subset \text{Equivalence} \subset \text{Adjunction}$$

containment sequence, for which

(79) $$\langle G, \mathrm{P}, \eta, \ni\rangle \in \text{Adjunction} \sim \text{Equivalence}\ .$$

Coda

13.14 The Art of Decoding Relational biology is “mathematical models seeking realization”. Given a model with the encoding functor

(80) $$\varepsilon : \langle N, \kappa(N) \rangle \to \langle M, \kappa(M) \rangle,$$

the task is to obtain its generalized inverse, the decoding functor

(81) $$\delta : \langle M, \kappa(M) \rangle \to \langle N, \kappa(N) \rangle.$$

The most efficient and formulaic optimization procedure (cf. Sections 10.12 and 10.15) turns out to be adjunction, which completes the commutative diagram of the modelling relation thus:

(82) 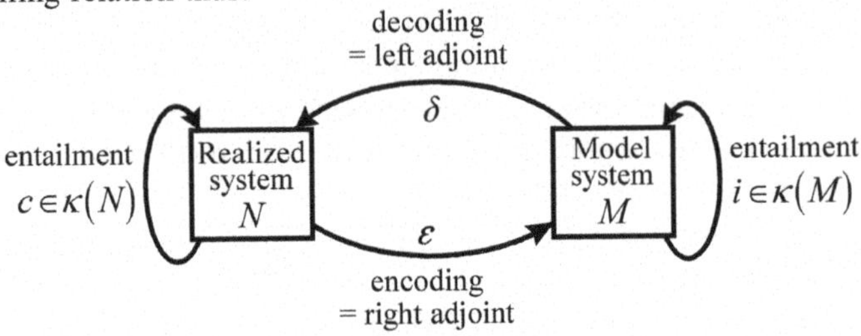

The approach to the study of life that is relational biology began with Nicolas Rashevsky’s 1954 paper *Topology and Life*, the subtitle of which is ‘In search of general mathematical principles in biology and sociology’. Here in *IL*, we have travelled a long road on the path of invertibility, culminating in the functorial connections of adjoints. Succinctly expressed entirely symbolically, then, *relational biology is*

(83) $$\delta \dashv \varepsilon.$$

We have just seen many examples of this general mathematical principle of adjunction. But it is important to note that one often only has *existence theorems* of left adjoints without specific means of *construction*. When one speaks of knowing some object ‘explicitly’, one usually means knowing its elements. An element of an object is a map *into* it: in an observable (Definition 5.7) $f \in H(X, Y)$, what is observed are the values $f(x) \in Y$. Since the decoding functor δ is a left adjoint, whence

(84) $$\varphi_{A,X} : M(\delta X, A) \to N(X, \varepsilon A),$$

one knows about maps *out of* δX, but not maps *into* δX. This is why *explicit* descriptions of left adjoints δ are often hard to come by. This is why realization $\delta(M) \subset N$ is a difficult problem. It is a crucial challenge in relational biology, our crux to bear.

The constructive aspects of mathematics, and especially the limitations of 'constructivism', are enlightening areas of research. These have incredibly profound ramifications for biology, for the rest of science, for technology, and for old philosophical questions pertaining to existence and to creation. The investigation of their implications in relational biology is the next peak to be scaled.

13.15 A Tangible Exercise: Protein Folding The native state of a protein molecule may be represented as a space curve passing through its α-carbons, whence completely characterized by two continuous mappings of the arc-length along the backbone, the curvature κ and the torsion τ, or equivalently the complex mapping $\kappa + i\tau$ [Louie & Somorjai, 1982, 1983, 1984]. If the protein is of length n (i.e., a polypeptide of n amino acids), then its three-dimensional structure in the curvature–torsion representation is a mapping in $A \in \mathbb{C}^n = \mathbf{Set}(n, \mathbb{C})$, or equivalently a finite sequence of complex numbers. Thus, geometrically, a protein molecule is an element of the sequence space c_{00}, the space of all infinite sequences with only a finite number of non-zero terms (sequences with finite support).

The primary structure of a protein of length n is an n-letter word $X \in M^n$ over the 20-letter alphabet M of amino acids, whence a member of the (countably infinite) free monoid M^* over 20 letters (cf. Section 12.9).

The mapping F that sends the tertiary⊕secondary structure $A \in \mathbb{C}^n \subset c_{00}$ of a protein to its primary structure $FA \in M^n \subset M^*$ 'forgets' its geometry. The recovery, which is precisely the exercise of the protein folding problem, is the construction of the left adjoint in the adjunction

(85) $$c_{00} \underset{F}{\overset{G}{\rightleftarrows}} M^*,$$

with bijections

(86) $$\varphi_{A,X} : c_{00}(GX, \mathbb{C}^n) \cong M^*(X, M^n).$$

In sum, the left adjoint functor folds proteins. Given a primary structure $X \in M^n \subset M^*$, the protein is folded with $G : X \mapsto GX \in \mathbb{C}^n \subset c_{00}$. I leave it as an exercise to the reader to explicitly construct the functor $G : M^* \to c_{00}$.

13.16 Keep Swinging

> Cultural heritage does not end at monuments and collections of objects. It also includes traditions or living expressions inherited from our ancestors and passed on to our descendants, such as oral traditions, performing arts, social practices, rituals, festive events, knowledge and practices concerning nature and the universe or the knowledge and skills to produce traditional crafts. While these may not be tangible — they cannot be touched — they are a very important part of our cultural heritage. This is intangible cultural heritage, a living form of heritage which is continuously recreated and which evolves as we adapt our practices and traditions in response to our environment. It provides a sense of identity and belonging in relation to our own cultures. As the world changes, modernisation and mechanisation are part of this living process — in many cases they might even assist and promote creativity. However, people still play the key role in the creation and carrying forward of intangible cultural heritage.
>
> — *Infokit: Questions and Answers about Intangible Cultural Heritage* (2011)
> UNESCO document 01855

What is life? The relational-biologic voyage, in the line from Rashevsky to Rosen to me, continues. I commit our future to the next generation, who will carry on what we have begun and journey to all the undiscovered countries. Μαθεῖν παθεῖν. In this sense, and also in its social sense, life itself is Purgatory. But Purgatory is precisely a projection from the present onto the future. This anticipation of an immanent and imminent future is why, as the French Romantic writer François-René de Chateaubriand has noted, *Purgatorio* surpasses *Paradiso* and *Inferno* in poetry. The world is a sojourn of all things. Time is a transversal of all ages. The essence of life is change.

Beati qui non tetigerunt et crediderunt. Euouae.

Acknowledgments

In pectore.

A. H. Louie
27 June, 2017

A. H. Louie, *Intangible Life*, Anticipation Science 2,
DOI 10.1007/978-3-319-65409-6

$$\underset{\varepsilon}{\overset{\delta}{\leftarrows\!\!\!\!\!\!\!\rightarrows}}$$

Bibliography

Degen, W. (2000). Pigeonhole and choice principles. *Mathematical Logic Quarterly, 46,* 313–334.

Hadlock, C. R. (1978). *Field theory and its classical problems*. Washington, DC: Mathematical Association of America.

Halmos, P. R. (1960). *Naive set theory*. Princeton, NJ: Van Nostrand.

Hofmeyr, J. H. S. (2017). Basic biological anticipation. In R. Poli (Ed.), *Handbook of anticipation*. New York: Springer. doi:10.1007/978-3-319-31737-3_51-1.

Klir, G. J. (2010). Book review: *More than life itself* by A.H. Louie. *International Journal of General Systems, 39,* 793–796.

Lawvere, W. (1969). Adjointness in foundations. *Dialectica, 23,* 281–296.

Lee, B. (1997). *Jeet Kune Do. Bruce Lee's commentaries on the martial way*. In J. Little (Ed.). Boston: Tuttle.

Louie, A. H. (1985). Categorical system theory. In R. Rosen (Ed.), *Theoretical biology and complexity: Three essays on the natural philosophy of complex systems* (pp. 69–163). Orlando, FL: Academic Press.

Louie, A. H. (2009, *ML*). *More than life itself: A synthetic continuation in relational biology*. Frankfurt: Ontos.

Louie, A. H. (2011). Essays on more than life itself. *Axiomathes, 21,* 473–489.

Louie, A. H. (2012). Explications of functional entailment in relational pathophysiology. *Axiomathes, 23,* 81–107.

Louie, A. H. (2013, *RL*). *The reflection of life: Functional entailment and imminence in relational biology*. New York: Springer.

Louie, A. H. (2017a). Relational biology. In R. Poli (Ed.), *Handbook of anticipation*. New York: Springer. doi:10.1007/978-3-319-31737-3_17-1.

Louie, A. H. (2017b). Mathematical foundations of anticipatory systems. In R. Poli (Ed.), *Handbook of anticipation*. New York: Springer. doi:10.1007/978-3-319-31737-3_21-1.

Louie, A. H., & Poli, R. (2011). The spread of hierarchical cycles. *International Journal of General Systems, 40,* 237–261.

Louie, A. H., & Poli, R. (2017). Complex systems. In R. Poli (Ed.), *Handbook of anticipation*. New York: Springer. doi:10.1007/978-3-319-31737-3_3-1.

A. H. Louie, *Intangible Life*, Anticipation Science 2,
DOI 10.1007/978-3-319-65409-6

Louie, A. H., & Somorjai, R. L. (1982). Differential geometry of proteins: A structural and dynamical representation of patterns. *Journal of Theoretical Biology, 98,* 189–209.

Louie, A. H., & Somorjai, R. L. (1983). Differential geometry of proteins: Helical approximations. *Journal of Molecular Biology, 168,* 143–162.

Louie, A. H., & Somorjai, R. L. (1984). Stieltjes integration and differential geometry: A model for enzyme recognition, discrimination, and catalysis. *Bulletin of Mathematical Biology, 46,* 745–764.

Mac Lane, S. (1997). *Category theory for the working mathematician* (2nd ed.). New York: Springer.

Nadin, M. (2015). Book review: Quo vadis relational biology? *The reflection of Life* by A.H. Louie *International Journal of General Systems*, 44, 111–120.

Noble, D. (2006). *The music of life: Biology beyond the genome*. Oxford: Oxford University Press.

Noble, D. (2017). *Dance to the tune of life: Biological relativity*. Cambridge: Cambridge University Press.

Ore, Ø. (1944). Galois connexions. *Transactions of the American Mathematical Society*, *55*, 493–513.

Poli, R. (Ed.). (2011). Special issue: Essays on *More than life itself. Axiomathes*, *21*(3), 373–489.

Rashevsky, N. (1954). Topology and life: In search of general mathematical principles in biology and sociology. *Bulletin of Mathematical Biophysics, 16,* 317–348.

Rashevsky, N. (1961). General discussion following the presentation of Robert Rosen's paper 'Recent progress in relational biology'. In H. L. Lucas (Ed.), *Proceedings of the Cullowhee conference on training in biomathematics* (pp. 205–233). Raleigh, NC: Typing Service.

Rosen, R. (1958a). A relational theory of biological systems. *Bulletin of Mathematical Biophysics, 20,* 245–260.

Rosen, R. (1958b). The representation of biological systems from the standpoint of the theory of categories. *Bulletin of Mathematical Biophysics, 20,* 317–341.

Rosen, R. (1959). On a logical paradox implicit in the notion of a self-reproducing automata. *Bulletin of Mathematical Biophysics, 21,* 387–394.

Rosen, R. (1962). Letter to the editor: Self-reproducing automaton. *Bulletin of Mathematical Biophysics, 24,* 243–245.

Rosen, R. (1972). Some relational cell models: The metabolism-repair systems. In R. Rosen (Ed.), *Foundations of Mathematical Biology* (Vol. 2, pp. 217–253). New York: Academic Press.

Rosen, R. (1974a). On the design of stable and reliable institutions. *International Journal of General Systems, 1,* 61–66.

Rosen, R. (1974b). Planning, management, policies and strategies: Four fuzzy concepts. *International Journal of General Systems, 1,* 245–252.

Rosen, R. (1985). *Anticipatory systems: Philosophical, mathematical, and methodological foundations*. Oxford: Pergamon Press.

Rosen, R. (1991). *Life Itself: A comprehensive inquiry into the nature, origin, and fabrication of life*. New York: Columbia University Press.

Rudin, W. (1986). *Real and complex analysis* (3rd ed.). New York: McGraw-Hill.

Schwartzman, S. (1994). *The words of mathematics: An etymological dictionary of mathematical terms used in English*. Washington DC: The Mathematical Association of America.

UNESCO. (1995). *Our creative diversity*. UNESCO document 101651, summary document 105586. Paris: UNESCO.

UNESCO. (2003). *Convention for the safeguarding of the intangible cultural heritage*. UNESCO Document 132540. Paris: UNESCO.

UNESCO. (2011). *Infokit: Questions and answers about intangible cultural heritage*. UNESCO document 01855. Paris: UNESCO.

Index

A. H. Louie, *Intangible Life*, Anticipation Science 2,
DOI 10.1007/978-3-319-65409-6

The manufacturer's authorised representative in the EU is Springer Nature Customer Service Centre GmbH, Europaplatz 3, 69115 Heidelberg, Germany. If you have any concerns regarding our products, please contact ProductSafety@springernature.com

Printed and bound by CPI Group (UK) Ltd, Croydon, CR0 4YY
25/08/2026
02203151-0003